Liaison Functions 2018

Core Programming Area at the 2018 AIChE Spring Meeting and 14th Global Congress of Process Safety

Orlando, Florida, USA
22 - 25 April 2018

ISBN: 978-1-5108-6448-1

Printed from e-media with permission by:

Curran Associates, Inc.
57 Morehouse Lane
Red Hook, NY 12571

Some format issues inherent in the e-media version may also appear in this print version.

Printed by Curran Associates, Inc. (2018)

For permission requests, please contact AIChE
at the address below.

AIChE
120 Wall Street, FL 23
New York, NY 10005-4020

Phone: (800) 242-4363
Fax: (203) 775-5177

www.aiche.org

Additional copies of this publication are available from:

Curran Associates, Inc.
57 Morehouse Lane
Red Hook, NY 12571 USA
Phone: 845-758-0400
Fax: 845-758-2633
Email: curran@proceedings.com
Web: www.proceedings.com

TABLE OF CONTENTS

517723 Building a Winning Culture: Harnessing the Power of Values

Abstract:

When The Chemours Company began in 2015 as a spinoff of DuPont, it faced daunting challenges, financial and otherwise. Some analysts contended it was doomed to fail. Instead, barely two years later it became a *Fortune 500* company and was recognized as *Chemistry and Engineering Newsâ€™* 2017 Company of the Year. Mark Vergnano will address the corporate values, including Safety Obsession, that have landed Chemours on the path to success.

512742 Be Prepared for Column Inspections

Abstract:

Distillation column inspections are an essential step to a successful installation or revamp. Safety, mechanical design and knowledge of the process intended for the column are just a few of the topics that must be covered before entering the column. Setting up a checklist of items to be checked alone often requires significant attention to detail. This presentation will provide an introduction to preparation for a safe and successful column inspection.

513323 Acid Gas Absorption Fundamentals

Abstract:

Absorption is a standard unit operation that has been studied by chemical engineering students for decades. One of its primary uses is for the removal of acid gases, like CO_2 and H_2S, from natural gas streams. Historically, aqueous amines and hot carbonate solutions have been used to perform these absorptions. Many different amines, solution strengths, activators and passivators have been tested and employed industrially. Approximately half of the industrial towers are trayed; half, packed. Foaming adversely affects the hydraulic performance in these systems.

Acid gas absorption is receiving appreciable new attention as concerns rise regarding CO_2 in the atmosphere. In addition, the United States has recently experienced significant growth opportunities in shale gas production which require improvements in gas treating technologies. This presentation reviews acid gas absorption and regeneration fundamentals. Several PFDâ€™s will be shown to illustrate important concepts. New absorbent solutions will be discussed, including ionic liquids. Flue gas applications will be discussed, including the difficulties associated with column pressure drops and very large column diameters. Rate-based modeling approaches will also be reviewed.

513341 Using Radioisotope Techniques for Process Investigations

Abstract:

In all chemical plants and refineries there generally is not enough instrumentation in the right places in our processes. During the design phase, decisions were made between knowing more about the process, which means buying more instruments, and saving capital funds. When a problem appears with the operation of a vessel, there are always gaps in the knowledge of what is happening due to a lack of instruments in the places where knowledge is not routinely required. Process engineers and operations personnel need to know what is happening so they can take corrective action. When knowledge is missing, they often take actions that make matters worse.

For distillation and separation processes Chemical Engineers have historically relied upon plant process measurements such as flow rates, temperatures, and pressures and model or simulation results for data with which to solve troubleshooting or re-design problems. An array of on-line diagnostic services are available that can provide additional data offering real-time information on how pieces of process equipment are actually operating. The most common application is Gamma Scanning. This test is primarily applied, but not necessarily limited to, distillation or separation columns. Gamma scans provide a density profile of the internal process of operating distillation columns and other process vessels. The density profile can be used to diagnose the hydraulic operating conditions of mass transfer devices such as damage to internals, flooding, degree of entrainment or weeping, liquid levels on trays and distributors, liquid distribution through packed beds, etc. The presenter will show case studies where scanning revealed vital process information that helped solve an operating problem, or helped make a revamp/re-design successful. Additional derivatives from gamma scanning of detecting flow distribution patterns will be demonstrated. The ThruVision Scan technique is a specialized horizontal gamma scan used to generate a topographic profile of the internal cross-sectional density of process equipment. This profile is useful for the detailed study of liquid flow distribution through packed columns as well as having applications beyond distillation columns. A hybrid application involving tracers and scanning on fixed bed catalytic reactors will also be discussed.

Another technique involving radioactive materials is the Tracer Test, in which a radioactive tracer is introduced into the process for investigating the process flows. External radiation detectors are placed at strategically chosen locations outside vessels and piping to monitor the tracer flow through the process. Either liquid, gas or a solid tracer can be utilized depending on process conditions and the fluid to be studied. This tracer technique can identify liquid or gas maldistribution in fixed or fluidized bed reactors. The residence time through a tank or reactor can be measured and compared with the calculated residence time. Observed differences can relate directly to the degree of mixing or channeling inside the reactor. These tracer techniques have also been used in identifying leakage of heat exchangers and valves.

512775 Ethanol to Infrastructure Compatible (Jet) Fuels Via Ethanol Coupling Reactions

Abstract:

In recent years, the maturation of the bioethanol industry has led to a more readily available platform for renewable fuels. However, the presence of the â €œblendwallâ€ problem, as well as the shift in focus towards infrastructure compatible fuels has resulted in a need for processes that can utilize ethanol as a building block for more valuable fuels and chemical products. Ethanol condensation is a well-known chemistry that couples alcohols through an aldol condensation process to generate higher oxygenates. The higher oxygenates can be directly used as a chemical commodity or easily converted to infrastructure compatible fuels. To demonstrate the viability of infrastructure compatible fuels production, in particular jet fuel, from the ethanol coupling pathway the higher oxygenates generated from the ethanol coupling chemistry sent through a sequence of catalytic processing steps to produce the fuel fractions. The final jet fuel fraction was then analyzed using ASTM methods to compare the fuel properties with standard jet values. For jet fuel, ASTM standard for the freeze point is < -40 °C and the ethanol coupling generated jet fraction is < -70°C, which surpasses the freeze point requirement. ASTM requirement for the density at 15°C is between 0.775 - 0.840 kg/L and the ethanol coupling generated jet fraction density value (0.789 kg/L) falls in the ASTM required range. This talk will discuss the individual catalytic steps involved in generating the infrastructure compatible fuels via ethanol coupling reactions and the final fuel properties.

512887 Purification of Tire Pyrolytic Oil (TPO) and Its Comparison with Diesel Fuel

Abstract:

Due to growing population and rapid industrialization the global demand for industrial rubber products is increasing. This increase in demand for rubber products has led to the increase in global rubber production which has devastating environmental effects. Disposal of waste rubber products is becoming an environmental challenge due to its non-bio degradability. The majority of the waste rubber product is generated from the worn out, damaged or discarded automotive tires also known as scrap tire. Approximately 1.5 billion tires are produced each year which eventually end up in the waste stream. Scrap tire consists of organic matter that is often disposed illegally in open lands, ultimately creating a big pollution hazard to environment and hygiene. Pyrolysis is considered as a useful technique for recycling of scrap tire by which oil, gas, carbon black and steel wires are obtained. Pyrolytic oil is currently receiving renewed interest as an alternative to fossil diesel fuel. In the present work, crude TPO obtained from vacuum pyrolysis method was purified via de-sulfurization followed by simple distillation. Decolorization and deodorization of the oil was also done to make it suitable for the industry. 30-35% of purified oil (product) was obtained after processing. A comparison of different properties of oil is done before and after purification. Properties of the purified TPO are also compared with diesel fuel for its suitability in diesel engines.

519831 "You were born with Two Ears and One Mouth for a Reason...": The Art of Listening for Understanding *(A GROW University Workshop)*

Abstract:

"You were born with Two Ears and One Mouth for a Reason ...": The Art of Listening for Understanding *(A GROW University Workshop)*

We spend about 70-80% of our waking hours in some sort of communication, 45% of which is listening! However, we are often listening with the intent to break into the conversation with our reply rather than to first truly understand. This workshop was originally part of a grassroots professional development series at Dow (called GROW U) that took the concepts found in Stephen Covey's Seven Habits of Highly Effective People, but looked at them in fresh ways. The workshop is centered on the concept of empathic listening from the 5th habit and will help you discover how to be an effective listener at work, leading to understanding of the real problems that need to be addressed in order to drive towards effective solutions.

The session will also provide an overview of how a group of Dow women and interested men (with management support) initiated a monthly workshop series called ***GROW University*** focused on personal and professional development. The topics are chosen by the group and the content of each workshop is developed and presented by small teams. This grassroots training program elevates the entire organization through the highly interactive workshops (which are open to everyone), while providing opportunity for individuals to demonstrate leadership, teaching ability, and passion for their chosen topics. The ultimate goal is to engender a higher level of satisfaction and productivity within a diverse and inclusive community. This concept was originally launched at one Dow site and has now gone global, indicating the inherent need for these types of growth opportunities.

512153 Rapid Development of New Product Formulations through the Efficient Utilization of Company Data

Abstract:

The continual development of new materials or products is the lifeblood of most chemical companies. Companies are constantly trying to: (i) develop products with novel properties for use in new applications, (ii) improve the quality and cost of existing products, (iii) reformulate products using new materials, and (iv) consolidate product formulations using fewer starting materials. However, their general success in these endeavors has been poor. In a recent Accenture benchmarking report (2015) entitled "Why Low-Risk Innovation Is Costly" (http://www.accenture.com/us-en/Pages/insight-low-risk-innovation-costly.aspx), executives indicated that indeed innovation is extremely important. 93% of executives surveyed said their companies depend on innovation for long-term success, but only 18% believed that their company derived a competitive advantage from its innovation strategy, and most of the executives were less satisfied with their company's current innovation performance (2015) than they were in 2009 (previous benchmarking report).

Product development at most big companies is characterized by:

- A lot of trial and error and intuition-based experimentation
- Some use of DOE's – but usually only in a very local manner – no global overview
- No repository for the accumulated product development knowledge of the company
  - No easily accessible database of all past development data
  - No good database on raw material properties
  - No repository for past modeling efforts
  - No good way of sharing information among developers
    - Intuitive experience of individual researchers is often only way they carry knowledge forward.
    - Little of that information is effectively transferred to other developers.
  - No effective ways of rapidly finding formulations involving new raw materials
  - Processing conditions are often not included in development studies, but left to the scale-up group

There are basically 3 general degrees of freedom in developing, improving, or reformulating products: (i) the choice of raw materials or ingredients, (ii) the ratios in which to combine or react them (formulation or recipe) and (iii) the process conditions under which to manufacture them. These three degrees of freedom are highly interactive, making it important to consider them simultaneously.

In this presentation we discuss powerful multivariate latent variable data analysis methods for extracting information from diverse development databases (raw materials, processes, formulations and quality), for optimally augmenting the databases with multivariate DOE's, and for using that information via optimization to rapidly formulate products that meet all the desired quality specifications at minimum cost.

A number of past development projects will be used to illustrate the approach. These include the development of novel functional polymers for use in medical devices and for the cores of the very popular Srixon golf balls (used by a large number of the touring pros), the development of novel high performance polymeric coatings, and the development of new food formulations.

Although this presentation will refer to advanced multivariate data analysis, design of experiments, and optimization, it will not be essential for the attendees to have a detailed knowledge of these in order to understand the essential ideas and the potential of the methods. The emphasis will be on the basic problem of rapid product development through the efficient quantitative utilization of company data, and on illustrating this approach with successful industrial examples.

513514 Innovating Innovation Processes for Chemical R&D By Treating As Metaphorical Chemical Processes

Abstract:

Both industrial and academic research produce innovations that can be used to add value to products and services. Many labs seek ways to produce innovations that 1) have higher value, 2) are faster to market, and 3) are produced as a consistent stream. We have found that by treating the innovation process itself like a chemical process, the flowrate of valuable innovations can be increased. In this talk we analyze the innovation process according to streams (e.g., desired outputs, inputs, recycles, purges) and unit operations (e.g., idea generation "reactors" and idea evaluation "separators"). This perspective enables the analysis and measurement of R&D, providing a coherent philosophy for innovation processes.

513106 An Experimental and Computational Study of the Four Blade Loop Impeller for Liquid Dispersion during Suspension Polymerization

Abstract:

The current work describes the characteristics of a Four Blade Loop Impeller (FBLI), which is intended for applications such as suspension polymerization, wherein a narrow range of suspended droplet diameters is desired. Using a dilute dispersion of silicone oil in water, an experimental measurement of the droplet size is performed using Particle Vision Microscopy (PVM) technique. Correlations between the maximum droplet chord length and maximum energy dissipation rate in the system are reported, in agreement with literature. The FBLI shows a narrower range of droplet sizes compared to a standard Pitched Blade turbine at the same impeller based Reynolds number. Computational Fluid Dynamics (CFD) simulations of this laboratory scale system with this impeller and other canonical paddle impellers showed a narrow rate of shear on the impeller surface for the FBLI. The CFD model was validated with experimental results from Particle Image Velocimetry measurements, and shaft power measurement. Power and flow numbers of the impeller are reported.

DIFREX^R Nexgen Expands and Advances Reactor & Technology Solutions

Subhash Dutta, Stephen Arnold, Aashish Gaurav, Jim Brenner
DIFREX LLC, 8900 Astronaut Blvd. Cape Canaveral, FL 32920 USA

Chemical reactors are the most vital elements of the chemical industry, contributing significantly to the total plant cost, efficiency of production, pollution control, number of separation steps needed downstream, and therefore to its ultimate profitability. Although the reactor is the heart of most process plants, it usually is treated as a "black box" or a proprietary item and is not covered by commercial simulators. Each technology developer or licensor uses its own procedure to develop its reactor model. Such a procedure often is lengthy and expensive, due to ill-defined steps, many trial-and-error mistakes, and excessive pilot-plant campaigns. Finally, even if the reactor is "successfully" scaled up to commercial size, the credibility of the design and the optimum operating conditions of the reactor often are questionable. "On the other hand, a plant owner may not have the know-how or confidence to modify, revamp, or modernize the reactor or its operation. The owner needs to build an in-house model for any such effort"[1]. The cover article "Build Robust Reactor Models" in the October 2000 issue of Chemical Engineering Progress presented a "well developed and tested procedure for building a robust reactor model/ model package, [with] tips and traps [plus] proven ways to substantially cut the cost and time for the effort"[1].

One of the authors of that article (Subhash Dutta) has spent his career on this work, plus other activities successfully "serving clients worldwide with the fastest, assured and most comprehensive reactor services and solutions in industry, to minimize risk and cost of process development; and to provide help in starting and making confident decisions on new reactor/technology or change/revamp of existing ones"[2]. Dutta's work has continued and advanced this work, including the General Reactor Model (GRM™) referenced in the article.

In addition to his further assistance to clients, he has formed a company, DIFREX^R LLC, which uses the GRM™ and also the wide range of know-how and expertise of an expanded team. Today DIFREX^R LLC is a team of process engineers based in Cape Canaveral, Florida, that provides reactor services and solutions to clients worldwide in areas such as design and consulting on new/existing reactor systems, critical review/quick-check of third-party designs, experimental development and validation of proposed and recommended designs, debottlenecking/retrofit/revamp, and emergency response[2].

This paper will provide more description of the GRM™ and the other services that DIFREX^R provides, including examples of past applications and references[1-34] and how they have contributed to faster development, evaluation, decision-making, etc., reducing costs and time, and have also led to increased standardization and enhanced certainty of the results, with additional economic benefits. These advances have been applied to a wide variety of reactor systems, with or without catalyst, and will be extended to more in the future.

1. Who says that a Simple Fixed Bed is the Right Choice?

Packed/Fixed Bed Reactors (PKB/FXB), which employ larger particles in a static bed, are traditionally viewed as the most *obvious or first choices* in industry partly due to the traditional notion that they are the easiest to scale up. In fact, the scale-up of a multi-tubular fixed bed (**TFXB**) system is not a scale-up at all. A commercial reactor of this system simply uses a multiple of the successfully performance-demonstrated single catalyst-filled tube to reach the commercial production goal. However, this class of reactors suffers from many serious drawbacks that include the following: a) high heat and mass transfer resistances leading to a steep axial temperature profile in the reactor, b) poor catalyst utilization or low catalyst effectiveness factor, due to its use of relatively large particles, c) possible maldistribution of gas between all tubes in multi-tubular design (including impact of catalyst not loaded equally), d) no continuous solid (catalyst or material) regeneration and replacement capability, e) no load-following capability, f) hot spot formations possibly leading to catalyst activity/selectivity loss and even catalyst/material vitrification, g) higher pressure drop, and h) restricted window of allowable feed composition, temperature and/or pressure to avoid possible temperature runaway for highly exothermic reactions.

However, there are many alternative reactor designs, as illustrated in **Figure 1**.

Fluidized Bed Reactors, both Bubbling/Turbulent (BFB) and circulating (CFB), do not suffer from the drawbacks of PKB/FXB reactors indicated above. The smaller particle sizes in these particulate systems lead to better heat and mass transfer rates, allowing operation at uniform bed temperature, which is normally preferred. Close to 100% catalyst utilization can be achieved, with no gas maldistribution. These reactors also provide continuous catalyst/material regeneration and replacement capability due to fluidizability of the chosen particulates that can be continuously discharged and recharged after regeneration and/or feeding a make-up stream. These reactors also have load-following capability, allowing operation across a broad window of production capacity as desired. They can be operated with lower pressure drops and are usually also free of hot spots. The windows of allowable feed composition, pressure and temperature are much wider compared with a fixed bed reactor, with less chance of reactor temperature runaway. On the other hand, the key drawbacks of a fluidized bed reactor include the following: a) potential dust emission with possible environmental concern, b) loss of expensive catalyst by entrainment, to the extent that fines are not fully captured and returned, c) particulate attrition causing excessive fines buildup, leading to excessive solid loss or depletion of fines that are needed for good fluidization, and other potential problems like line plugging and agglomeration, d) potential erosion of lines or internal heat transfer tubes, e) poorer solid-gas mass and heat transfer rates for BFB through the interface films of gas bubbles or voids, and f) potential neglect of safety concerns (even unacceptable bypassing of safety instrumentation and procedures) if operators do not fully understand, and become complacent with respect to the high flame-arresting safety benefits of a BFB, e.g., for partial oxidation (POX) reactions, including knowing when or where they do not apply.

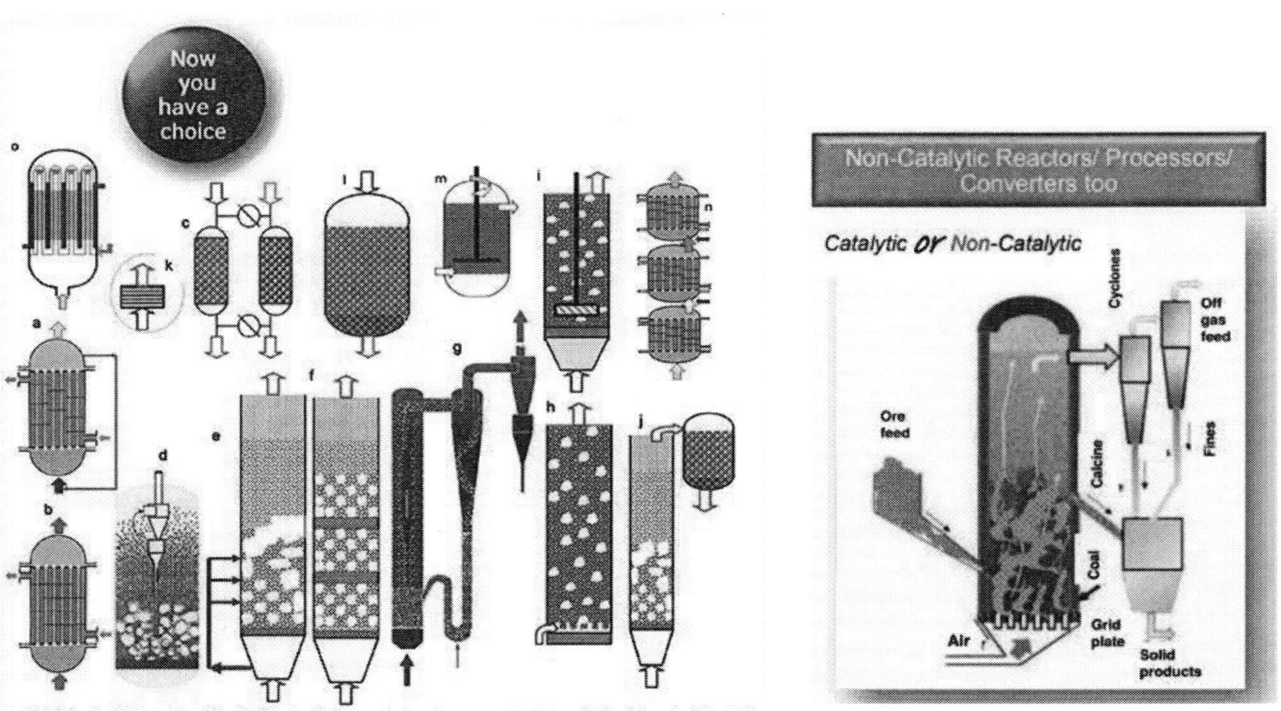

a, b)Tubular fixed bed reactor with and without product recycle: c)reactor-regenerator dual-packed-bed: d, e, f)bubbling fluid bed without and with quench/ inter-stage feed injection, and with redistributors: g)fast/ circulating fluid bed: h, i)fluidized bubble-column and stirred slurry reactor: j)hybrid (fluidized bed+polishing) reactor: k)micro-channel reactor, l)packed bed, m)stirred tank, n) quenched tubular fixed bed, o)double wall bayonet tube reactor

Figure 1: DIFREX technologies encompass numerous catalytic and non-catalytic reactors

Circulating Fluidized Bed Reactors (CFBs) are usually designed for fast reactions and for reactor-regenerator dual-bed systems. In the dual-bed system, the solid particulates, catalyst or non-catalytic material, can be regenerated in the regenerator bed, or sometimes in a transfer line, and be returned back to the reactor in a continuous circulation of the solid. (This is also done with some BFBs.) The primary section of a CFB system operates without gas bubbles or voids, and has minimal solid-gas heat and mass transfer resistances. However, it suffers from the following drawbacks compared with the other systems: a) bed voidage throughout these reactors is usually very high (in the range of 88-98%), thus, solid concentration per unit bed volume very low, resulting in much lower reaction rate per unit bed volume, b) these reactors operate at relatively much higher gas velocities and therefore are applicable to fast reactions only, c) at the high gas velocities, which includes the catalyst movement, they demand much tougher or more attrition-resistant catalyst or material unless the material is desired to be broken down (or reacted away) and to pass once-through as in gasifiers or combustors, d) the reactors need to be much taller (in the order of 30 meters or even more) for adequate reaction residence time to achieve satisfactory conversions at their high gas velocities.

Figure 2 illustrates the extensive and convenient framework and easy use of the DIFREX software as represented as a **dial-a-reactor**. It incorporates detailed flow and hydrodynamic behavior for many different reactor types, allowing prompt and easy calculation switching (dialing) from one reactor type to another, for the same chemistry and kinetics of the reaction system. Better, faster initial determination of the best conceptual/preliminary design for a reaction system, catalyst or process can therefore ***now*** be achieved by comparing the design and performance details projected by the software at the earliest project stages. The best choice could be a packed bed (PKB), a tubular fixed bed (TFXB), a bubbling/turbulent fluid (BFB), a circulating fluid bed (CFB), all with or without quench and/or recycle, or a hybrid reactor (CSTR-CFB-BFB) as in a typical FCC reactor or other system. The best **Reactor Of Choice (ROC)** can now be determined along with the **Catalyst or Material Of Choice (COC/MOC)** and much earlier – even while the solid catalyst or material is being developed. Final determination of best ROC,

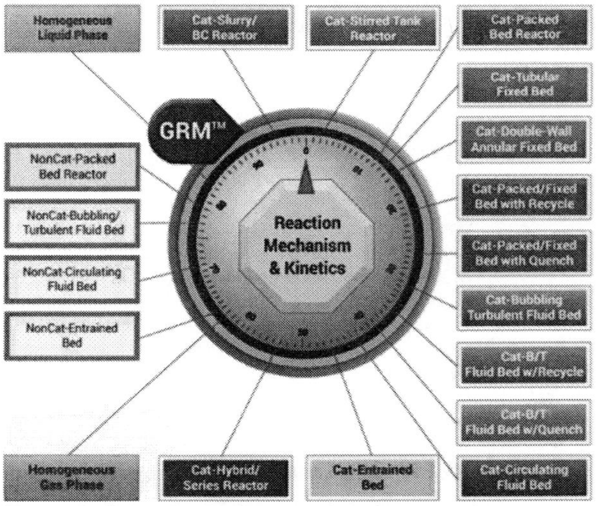

Figure 2: Extensive and convenient framework for reactor selection

however, would be determined by **consideration of other criteria** that include cost, particulate handling/recycle, strength and attrition resistance of solid catalyst, and miscellaneous other characteristics like particulate fluidizability and agglomeration problems for fluidized bed systems. However, the software allows one to reach each ***GO/NO-GO decision gate*** much sooner, leading to substantial savings in both cost and time for catalyst, process or technology development.

Current DIFREX team members were instrumental in successful commercialization of several new process technologies, involving all four basic solid-gas reactor types: PKB, TFXB, BFB and CFB reactors and their variations. A prime example is the ALMA process, a fluidized bed POX process for production of maleic anhydride from n-butane and air. These designs were products of seamless integration of particle technology (including particulate flow and hydrodynamic behaviors) with chemical reaction engineering principles. Results of some of these works are available in a series of publications by Dutta, Arnold, and other authors[1-20]. Many of the theories and procedures used in these works can be found in the two classical books on Fluidization Engineering and Chemical Reaction Engineering by Kunii and Levenspiel[21-22].

Scale-Up Fear: No More

Due to significant advances made in both chemical reaction engineering and particulate system design and engineering, and successful demonstrations over the past three decades in commercial processes, there is little reason any more to fear scale-up failure for most of the reactor types used in the industry today. Failures and/or delays still seen are primarily due to questionable or random (non-standardized) kinetic data generation procedures that lead to unreliable data that are not suitable for successful scale-up.

2. Who says that Plant Expansion or Upgrade/Change is Difficult?

DIFREX software, services and solutions integrate chemical reaction engineering with particle technology to provide the right solutions for design, development, scale-up and successful commercialization of gas-solid reactors, enabling one to promptly establish the design and performance of virtually any particle or solid-gas reactor type for reaction systems of different complexities. Since it has fundamental basis in the two engineering sciences (reaction and particle) and avoids much of the empiricism of traditional process development, especially in the HC processing industry, it is better suited for new designs and also revamp or upgrade of an existing technology/ reactor/ process, particularly for a new feed or product stream. This refers to both catalytic and non-catalytic solid-gas reactions, including the use of a catalyst (active components with or without carrier) as the solid for catalytic processes or a non-catalytic solid particle like coal, biomass or a mineral (FeO or NiO, for example) for combustors, gasifiers, and metal extraction/roasting processes, and also solid particles as adsorbents to transfer atoms and other molecular species from one stream or vessel to another.

Reach *Decision Gate* Quickly

*How **not** to see that moth-balled R&D/ Pilot Plant/ Demo Campaign Again?* Whether it is a new venture or a retrofit, revamp or upgrade, delay and cost overrun are common even to reach the first ***Go/No-Go decision gate***. This adds to the pervasive fear of unmatched (below promised/projected) reactor performance, or worse, total failure. We have witnessed these events in multiple ventures that succeeded or failed. We have critically examined and documented the causes and effects. We are now ready to offer solutions to minimize cost and time to reach the decision gates and then minimize fear and risk for the *final* development.

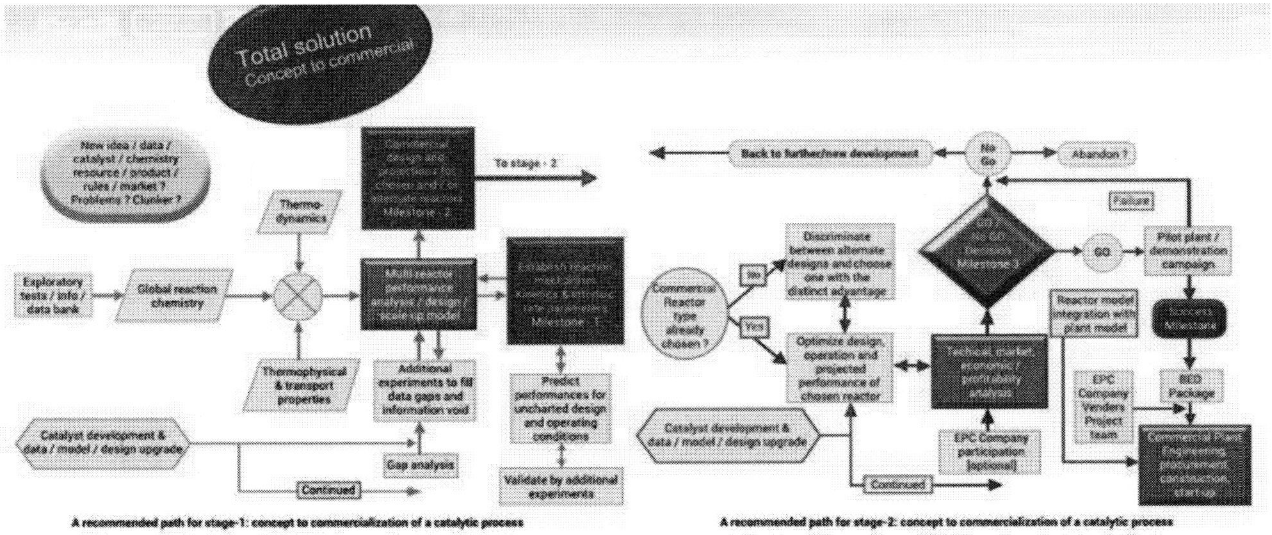

Figure 3: Concept to commercial action flowchart, critically examining reaction and process engineering at each step

A2Z: Fastest Path from Lab to Commercial

Due to combined background, experience and expertise in concept validation, reaction engineering, detailed design of reactors and auxiliary systems (like cyclones, diplegs, circulation loops, bed internals, bubble size control, distributor grids), detailed modeling and simulation, and process engineering, DIFREX can help in your next or ongoing venture. The service covers the entire process or ***total solution*** (**Figure 3**) from concept or lab to commercial, and accomplish the task successfully at the fastest speed possible.

Figure 4: GTL unit on a shipboard

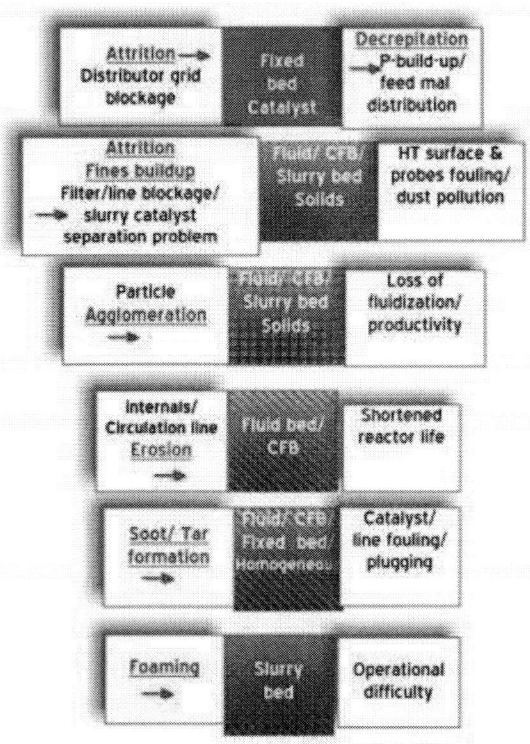

Figure 6: Prominent reactor design issues often ignored in most design software models

Go Ahead

So, go ahead. • Add that sulfur guard/device for emission control or for feed gas treatment, • Place it at a well mouth or along distribution lines, • Install a flare/ waste stream converter for profit and environment, • Install a mercury emission control for your sludge/waste incinerator, • Install concentration monitoring controls to meet environmental regulations, • Install a GTL unit at a well mouth, remote island/location or on shipboard (**Figure 4**), • Carry out *retrofit, revamp or upgrade* you have postponed, • Revive troubled/idled units, • Minimize over- or under-designs.

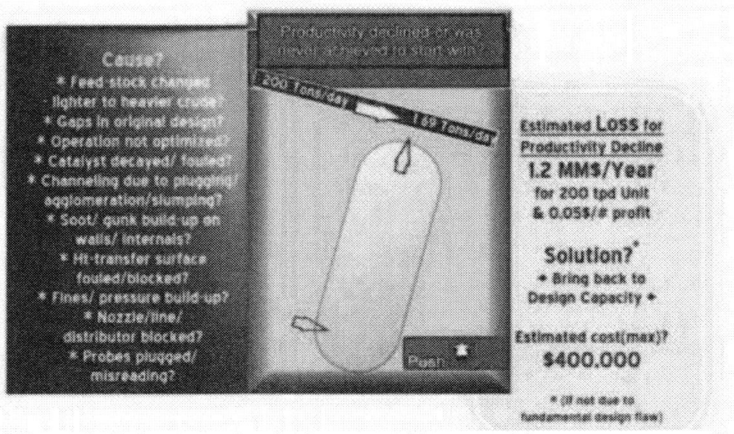

Figure 5: Robust reactor design can address problems in underperforming, uncontrollable, unpredictable reactor systems

Take the fastest path to 3-D Reactor Solution → Design, Development and Debottlenecking

Underperforming, uncontrollable, unpredictable, problem-ridden or dying reactor? Or, one sitting on the side-line/off-line? Help is NOW available. Both on- and off-site plant unit debottlenecking on short notice and in-depth assessment of estimated benefits, cost and return on recommendations or optimization of running unit or retrofit/revamp is available. Most often, the benefits and quick returns would strongly justify the modification as shown schematically in **Figure 5**. Prompt diagnosis and solutions to troubled or idled reactors are *now* available.

One-stop solution

Although many details beyond the key design specs and operating conditions are to be filled by process engineers, there are issues that still fall under the primary responsibility of the reactor designer. Some of these issues are shown in **Figure 6**. These aspects are usually beyond the scope of most computer models. But ignoring these issues in advance, or inadequate/wrong solutions to these anticipated problems/issues may lead not only to project failure, but also to disastrous consequence in a running plant.

Reactor as "unit operation": Unlock true potentials plus easy revamps and upgrades – plant-wide

A reactor or reaction system (shown in **Figure 7**) is the heart of the plant, but it is often not treated as another unit operation in a process plant – like a separator, settler, heat exchanger, pump or compressor. It is usually treated as a **black box** where all upstream operations feed into and downstream operations begin. The rest of the plant design utilizing unit operations is constrained to meet the needs of the feed and narrow feed/operating conditions and to *treat the effluents* according to the original design of the reactor. Reasons include protections of trade secrets and control by reactor vendors, and presumed risk and fear.

Figure 7: Reactor as heart of the plant, but often treated as black box

DIFREX technologies simplify the design and development of reaction systems as well as debottlenecking operations. One can now unlock the full potentials of the plant a) by treating each reactor as another *unit operation* and its *true/full optimization* by going beyond the *narrow* boundaries of designs and operations – possibly *locked in* by the original design and pre-conception, b) optimized for *new/changed feedstock/* feedstock composition, and c) optimized for new/changed product/ product composition/ market demand. These operations can lead to savings in upstream and/or downstream processing costs due to improved/new reactor feed and effluent conditions or better yield. Although the original capital cost of a reactor may be only 10-30% of total plant capital cost, a substantial savings in capital and/or operating costs of the up- and downstream processing units by the new and true or full plant optimization process may lead to substantially improved profit from the operating plant as shown schematically in **Figure 8**.

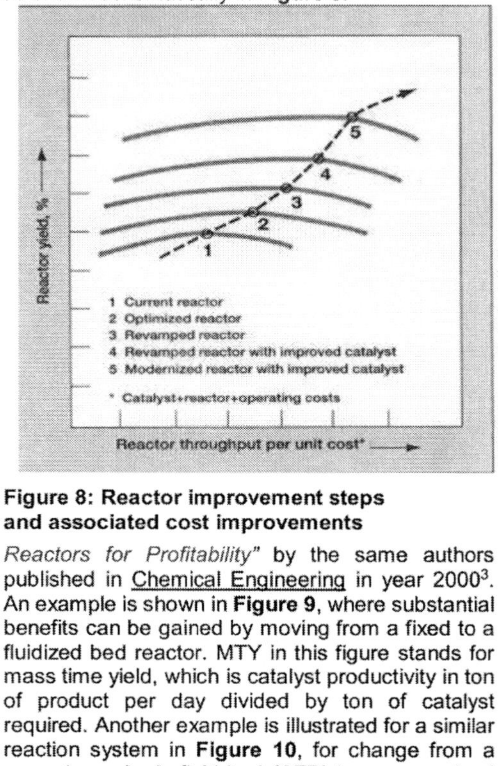

Figure 8: Reactor improvement steps and associated cost improvements

Plant Expansion or Upgrade/Change

Plant Expansion or Upgrade is *now* far easier and cost-effective, as illustrated in a series of three publications by Dutta and Gualy in Hydrocarbon Processing in 1999, titled *"General Reactor Model Improves HPI Applications"*, *"Overhaul Process Reactors"*, and *"Modernize Process Reactors"* [4-6], and in *"Super-modeling Reactors for Profitability"* by the same authors published in Chemical Engineering in year 2000[3]. An example is shown in **Figure 9**, where substantial benefits can be gained by moving from a fixed to a fluidized bed reactor. MTY in this figure stands for mass time yield, which is catalyst productivity in ton of product per day divided by ton of catalyst required. Another example is illustrated for a similar reaction system in **Figure 10**, for change from a near-atmospheric fluid bed (AFB) to a pressurized fluid bed (PFB) for substantial process benefits. Both changes were accomplished for maleic anhydride fluidized bed development and displacement of multitubular fixed-bed reactors for the ALMA Process, but with different actual parameter values and impacts than shown. Clearer views and explanations for these examples and more are available in the four cited references.

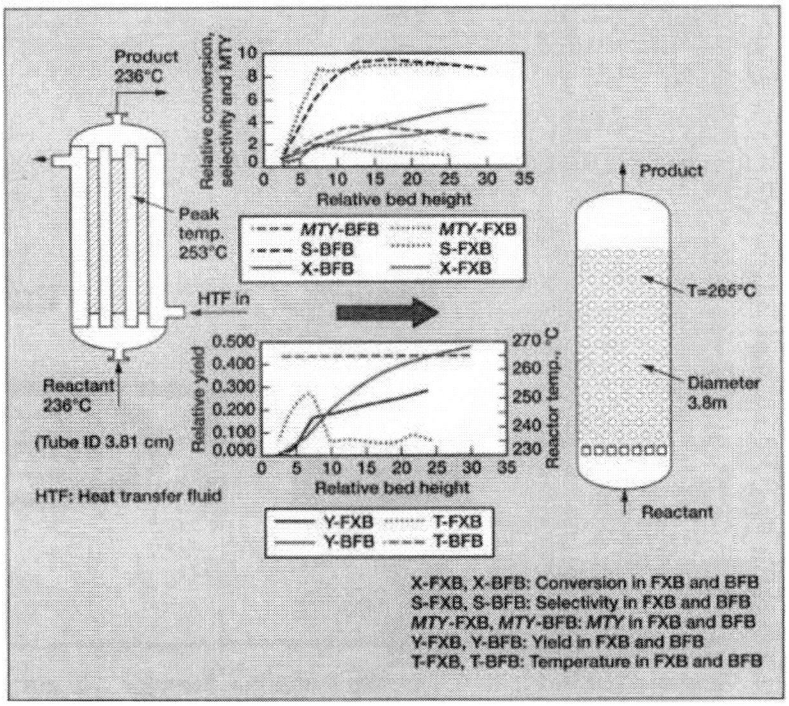

Figure 9: Expected benefits of a switch from Fixed Bed to Fluidized Bed reactor: illustrative example

For key hydrocarbon processing reactors, for example, for a typical FCC reactor, simulation and performance optimization/improvement can be more easily achieved by modeling it as a hybrid reactor system consisting of a CSTR at the bottom followed by a CFB and then a BFB on the top, as illustrated in **Figure 11**. Hydrocracking and isomerization reactions for heavy oil upgrade can be simulated by a reaction scheme shown in **Figure 12** using surrogate compounds. Detailed accounts of such treatment can be found on the DIFREX website[32] and in the references cited above.

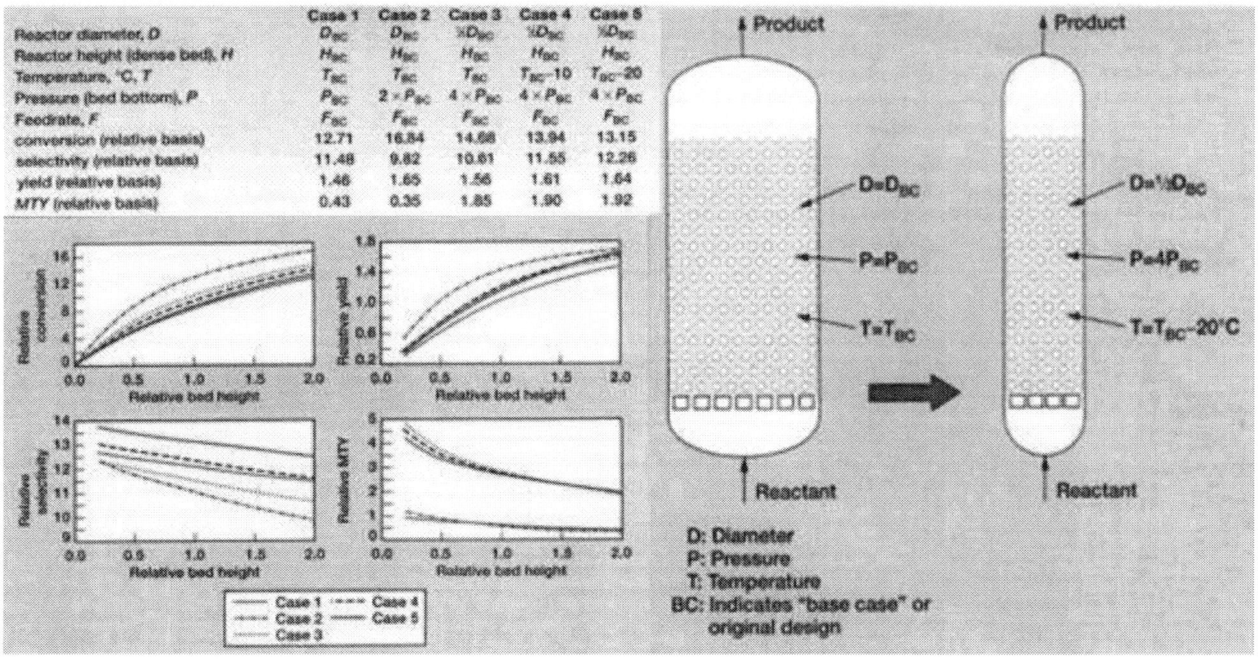

Figure 10: Switch from Atmospheric Fluid Bed to Pressurized Fluid Bed: illustrative example

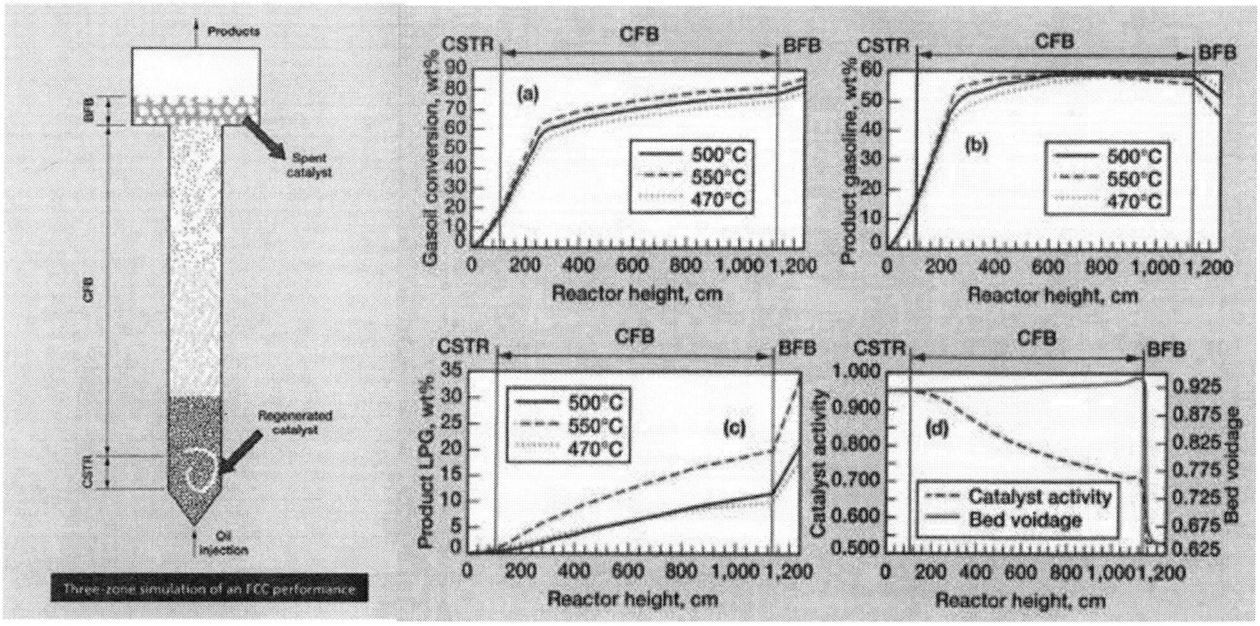

Figure 11: Performance Simulation of FCC with a simple 5-lump reaction model

3. Process Development: Randomness, Risk and Uncertainty

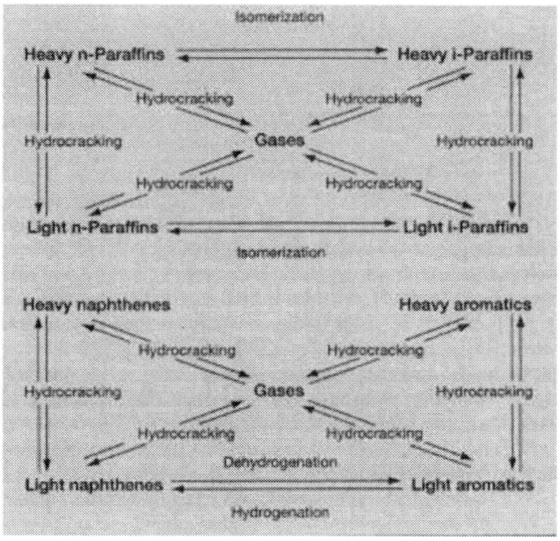

Figure 12: Typical Heavy Hydrocarbons Conversion reaction network

Standardized procedures:
For head start and minimum delay and failure

No standardized procedures or steps currently exist for any of the following reactor tasks: • upgrade/revamp, • servicing, • data/performance analysis, • catalysis design/development, • optimization, • catalyst/process development, • scale-up/design, • modeling, • operation/control and • process innovation. The entire field is currently run by individual choice, conflicting opinions, randomness, and many trial-and-error steps. This causes cost overruns, delays, and risk in a new venture and, in many cases, project failures. Standardized procedures/steps in carrying out one or more of these tasks from concept to commercialization for a new technology/ catalyst/ reactor or for upgrade/ revamp of an existing or old technology/ catalyst/ reactor are now offered by DIFREX for head start and to minimize project delays and possible failures. It is accomplished by minimizing the randomness of the very process of development, design or redesign/revamp efforts of many projects that proceed like a runaway reaction and become uncontrollable. DIFREX offers a specific roadmap, schedule and milestone chart customized for each project or subproject, which help both in cost containment and timely and successful completion, including • catalyst formulation, • test rigs design, • test plan and matrix, • performance and data analysis, • kinetic and reactor model development, • safe reactor design, and beyond. The standardized procedure is applicable to a new technology/ catalyst/ reactor, an operating plant reactor, and also for using the reactor as another "unit operation" in the plant as shown schematically in **Figure 13** and powered by a friendly General Reactor Model software "smartpack" (GRM™). This is coupled with additional offers/advantages and coverage as shown in **Chart 1**.

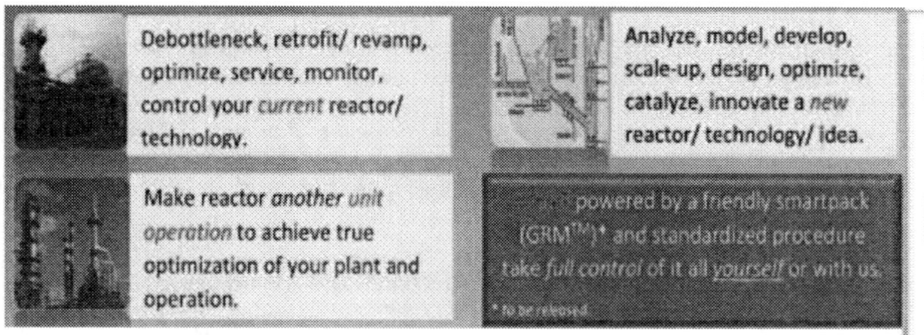

Figure 13: DIFREX offers standardized reactor design and implementation procedures

• *'No-Oops' A2Z solutions* NOW cover a *vista of reactors – reactors of virtually any kind* involving *reactions of any kind or complexity* and *even non-stoichiometric* reactions as in refineries, coal and biomass conversions.

• Tasks are *now* completed at a *fraction of time and cost* to reach *the Decision (Go-No Go)* Gate.

• Evaluate *alternate reactor concepts* at about same time to make *the right decision* on *ROC* (*Reactor of Choice*).

• *Head start* and *total CTC* (*concept-to-commercial*) services to *minimize failure*, and *improved reactor safety*.

• Aided by a *unique*, friendly, *secured and stand-alone* GRM™ software package (*smartpack*) that can hide in a wallet, and *standardized procedures*, tasks *now* become *far easier* than ever before.

• Building a *robust kinetic model* from *raw data* and minimum information that may be *key* to *commercial* success *and safe* operation is *now* made *easy*.

• Performance *guarantee* can be offered due to *our 'zero-failure' track record* of *multiple commercial successes* with fixed and fluidized and quenched *bed reactors* - backed by *our edge and insight* and advance 'oops' analysis.

• *Enterprise solutions* for sustained growth and edge.

Chart 1: Key/unique features of the services offered by DIFREX

Experimental Program

Misguided, lengthy and/or redundant and useless data generation often result from an experimental campaign for a new process development. DIFREX helps prioritize the areas of focus in recommending a test matrix that is conducive to collection of data most useful for promptly establishing reliable reaction mechanism and kinetics, catalyst screening, developing the reactor model, and successful scale-up and optimization of the commercial reactor design.

Kinetic Data, Temperature and Activation Energy (E)

DIFREX considers establishment of a reliable process/reaction kinetic model as the most important technical task for prompt and successful design, development or commercialization of a process. For catalytic gas-solid reactors, DIFREX recommends a specific (integral) test reactor design, plan and operating matrix for generation of *useful* kinetic data. Emphasis is placed primarily on measurement of the reactor axial temperature profile as accurately as possible, since the effect of temperature profile on the reactor or catalyst performance is usually far more than that of any other operating or design variable in most reactors. An accurate measurement of the temperature profile is necessary for evaluating the activation energy (E) of each reaction from the observed reactor performance data at various design and operating conditions. Relative values of the activation energies of reactions involved determine the product and by-product selectivities and yields for optimization for each reactor.

A reliable kinetic model must properly account for heat and/or mass transfer and other effects (shown schematically in **Figure 14**) that may mask the intrinsic reaction kinetics of a reaction/conversion process. Building a reliable kinetic model is a difficult and time-consuming task. DIFREX makes it easy even with minimal information in hand. DIFREX *has built reaction models as complex as in coal combustion, gasification and pyrolysis, hypergolic reactions of gel propellants in rocket engines and hydrocracking/hydrotreating of heavy oils, to as simple as ammonia and methanol syntheses.*

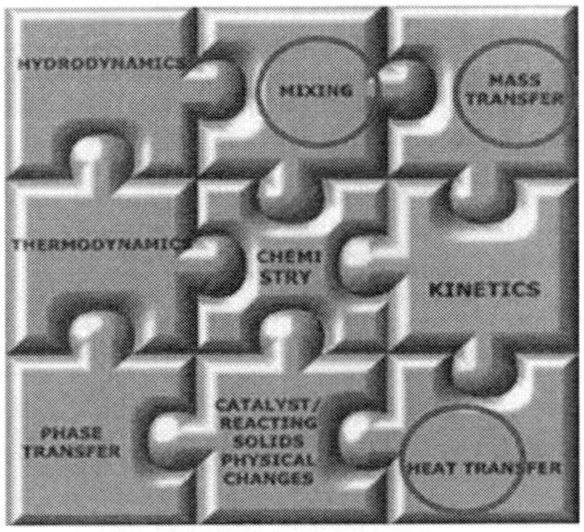

Figure 14: DIFREX kinetic models account for effects of other phenomena on intrinsic kinetics/process

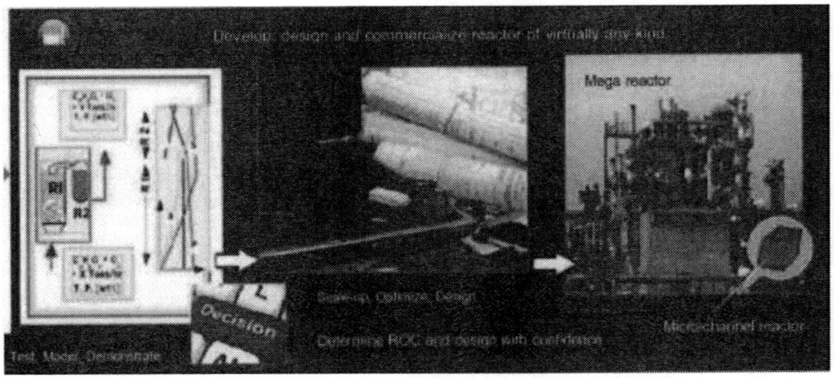

Figure 15: Typical progression from concept to commercialization via DECISION gate

4. Why MUST a New Technology Venture be Expensive?

Speedy and satisfactory solutions need more than a math-pack

Judgment is needed in *discrimination* between alternative information sources, methodologies, procedures, correlations and databases to create *effective* and *easy* solutions with *adequate details*. Theories, correlations and procedures need to be tested and *re-validated* at every scale from micro- to small- to bench- to pilot- to commercial reactor/reaction systems of *every kind – to minimize risk* of scale-up failure or for evaluating/switching from one to another reactor system. DIFREX offers *zero-failure multi-technology commercialization track record* and extensive *experience* dedicated *exclusively to reactors of virtually every kind* and backed by coverage and track-record *unmatched* in the field. Also DIFREX design is built on *five focal points* – ***ease***, ***choice***, ***agility***, ***safety***, performance ***ensurance***.

Time is everything

Now it is possible to move from concept to commercial (schematically shown in **Figure 15**) with confidence and at a much faster speed. The path includes catalyst/material development, determination of COC/MOC (Catalyst/Material Of Choice), testing and demonstration, modeling, design, scale-up, optimization, Go/No-Go decision, determination of ROC (Reactor Of Choice), process and detailed engineering, fabrication, construction/installation, commissioning and follow-up.

Both Catalytic and Non-Catalytic Reactor Designs

DIFREX's added strength is also in the design of non-catalytic (and mixed catalytic + non-catalytic) reactors wherein solids take part in both chemical and physical transformations. Areas covered include combustors/ incinerators/ gasifiers/ pyrolyzers of biomass/ sludge/ coal, minerals roasters/ calciners, desulfurizers and Hg removal units.

Design/ Development/ Revamp Done in 3 Quick and Easy Steps

It is accomplished by utilizing the GRM^{TM} Smartpack and following steps 1-2-3 (**Figure 16**). It is easy, prompt and user-friendly. Details are provided in references 32 and 33. Results for a comparison of design specifications and performances of three FT synthesis reaction systems are shown in **Fig. 17**, a quenched tubular fixed bed (QTFXB), tubular fixed bed (TFXB) and bubbling fluid bed (BFB). The *"Best Choice"* chosen here is from the comparison of the yields only. The final choice can be made by the user based on additional considerations that include mechanical/fabrication constraints, reactor cost, catalyst fines emission/management issues, attrition loss and circulation line erosion, among others.

The best Reactor Of Choice (ROC) for most systems can be determined ***within a week or two, if not in a couple of days***, provided that and when the back-up test/ performance data of the reaction system are sufficiently complete and reliable. Furthermore, the DIFREX database on reaction mechanism and kinetics, thermochemical and transport properties, and methodologies can help quickly bridge the gap between available and necessary information and leapfrog a new technology venture to success.

Step 1

Collect system performance data on COC/MOC (Catalyst/ Material of Choice). Establish process mechanism and kinetics following the guideline provided. If full information on process mechanism and kinetics not available, the SmartPack built-in database should help you to fill in most of it for many systems. Preliminary design based on it should be adequate for most applications.

Step 2

Define your process by answering five (5) questions on this page:
1. Objective
2. Feed
3. Key Product/ Product Group or Process
4. Reactor of Choice from the list below
5. Scale – Bench, Pilot/Demo or Commercial

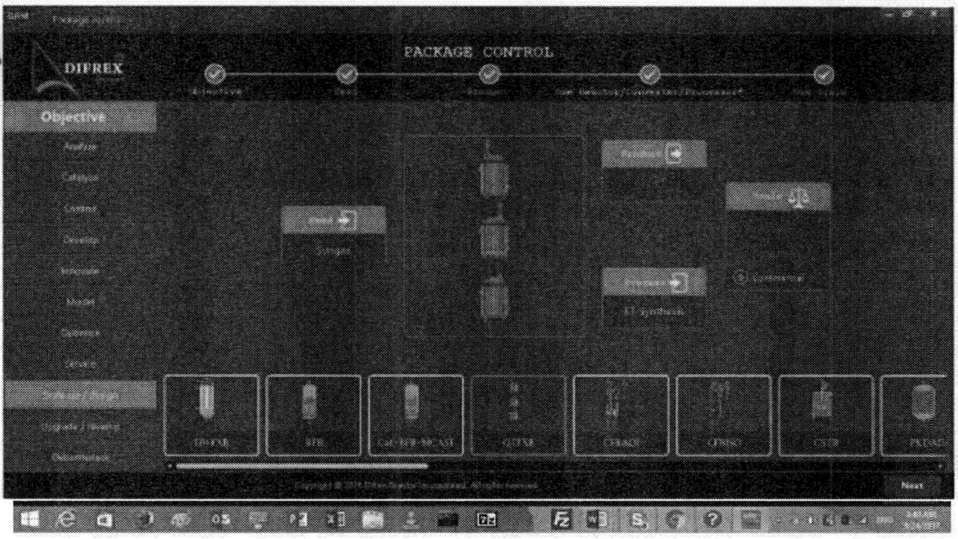

Step 3

Input* process mechanism/ kinetics, design/ operating conditions and catalyst/ material specs per top menu bar. Then:
1. Generate Input File
2. Execute

See input summary and results, showing complete design and performance summary in tabular and graphical forms and complete details® in 'more >>' sheets.
- Complete case studies for the chosen Reactor by changing one or more variables on on-screen Input table.
- Compare cases on-screen.
- Repeat for another Reactor.
- Pick the Best or RIGHT reactor.

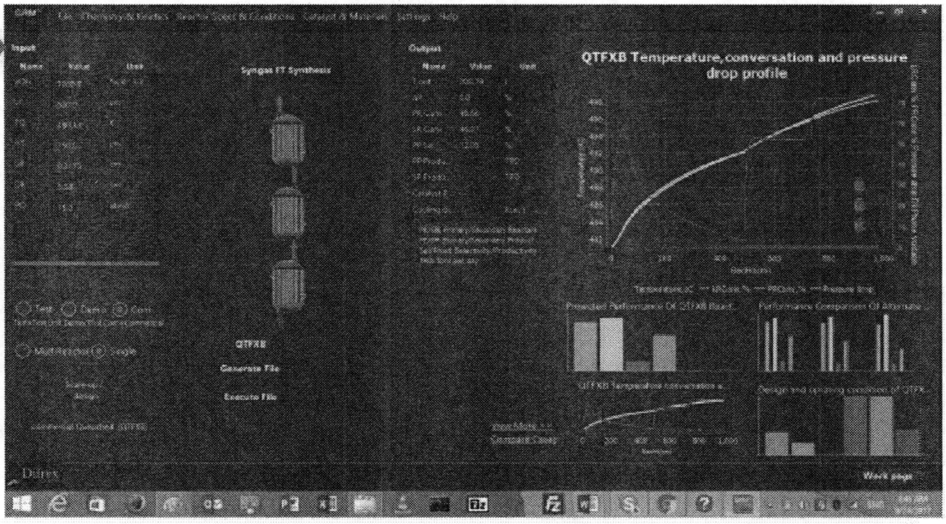

*Quick, friendly, easy and error-checked entries
®Complete temperature, pressure, composition, productivity, conversion and selectivity profiles, catalyst productivity and effectiveness factors, deactivation, equilibrium constants, reaction rates, cooling loads, thermo-chemical and transport property changes, heat and mass transfer modes, hydrodynamic behaviors like bubble growth, coalescence, slugging, grid, splash zone and free-board reactions, voidage profiles, TDH, solids loadings, flow regimes, entrainment rates, gas solid slip and more ....

Figure 16: Reactor design and development is as easy as 1-2-3 with DIFREX GRM™ software along with Reactor Of Choice selection

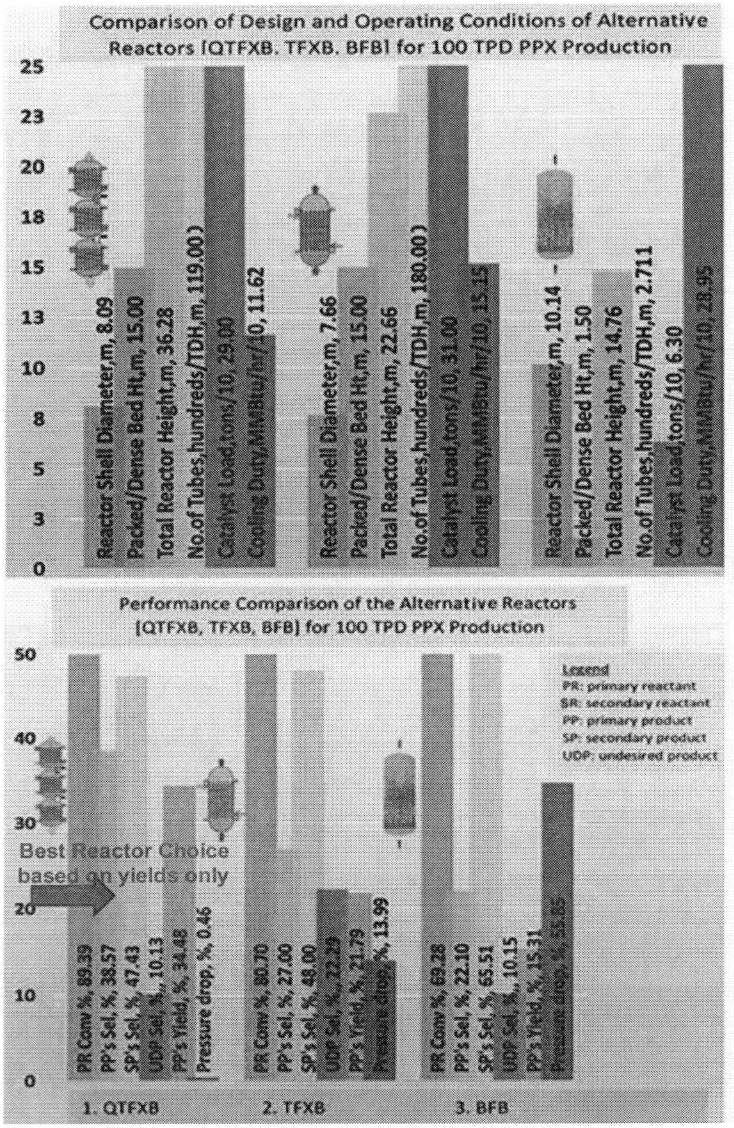

Figure 17: Comparison of design specs and performances of 3 reactors for an FT synthesis reaction system along with Reactor of Choice selection

Enterprise solution: Strategic/fast-track action and sustained growth

Fast-Track Technology: Helping to fast-track process and technology development is one of our core strengths. Our structured, industry-proven and 'no-oops' development strategy should get you to your goals faster. We did it before, Let us do it again ... for you.

Disruptive change is the new norm that makes sustained growth challenging. Reactors in your plant face the brunt of this challenge as feeds, product demands and markets can change suddenly. DIFREX can help in your strategy to be proactive not only to minimize losses due to such disruptions, but also to capitalize on them.

DIFREX can also help infuse that vibrant **culture** into your strategy at both administrative and functional levels that deliver the strategic advantage – by fueling the spirits (desires, engagement, and execution) for all tasks around your reactors including R&D. We help you identify a **strategic niche** and make your strategic decision possible by our in-depth and unbiased analysis of both market potentials and true technical positions of competing/emerging technologies.

Integrated Technology Development (ITD): A New Concept

Traditionally, process design or technology development begins ***after*** determination of the Catalyst or Material Of Choice (COC/MOC) is reached. This determination is mostly made by seeking a stable (non- or slowly deactivating) catalyst or material that is most appropriate for a fixed-bed reactor only. For a fluidized-bed reactor, where a catalyst or material can be regenerated, such a constraint does not exist. A fast-deactivating, but otherwise advantageous catalyst or material, with high activity and selectivity, is perfectly suitable for a fluidized bed system if these can be regenerated and recycled. Therefore, it is advisable to proceed with determination of the Reactor Of Choice (ROC) or the process technology ***much sooner*** in the technology development path. This is *now* possible and can be accomplished easily. In the traditional approach, a reactor has to match the demand of the Catalyst or Material Of Choice, which often leads to the use of a fixed bed as the sole or first Reactor Of Choice. This could be a poor choice when compared with alternative reactors. Using the simultaneous or Integrated Technology Development (ITD) approach, the right reactor and/or the technology can be determined much faster.

5. Future of Business – Now

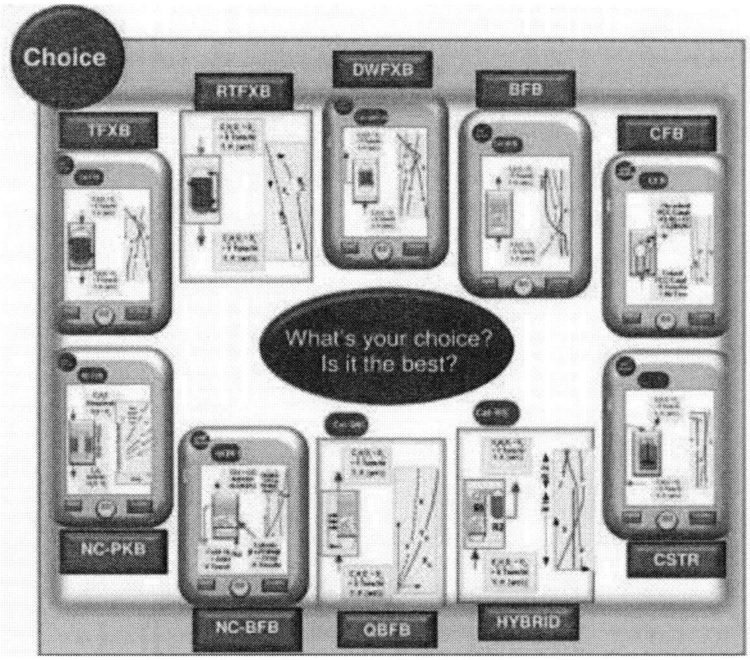

Figure 18: Alternative Designs and Development Paths now made easy

Process Technology Development will not be the same

It will be better, easier and much faster, leading to a quantum jump in *productivity* of design and development teams. Much of the task could be accomplished *quickly* in a hand-held device as shown schematically in **Figure 18** - with all security shields in place.

Micro-Channel Reactor (μ-CR) Technology [Process Intensification]:

For small to medium scale plants a micro-channel reactor (**Figure 19**) offers an advantageous alternative due to many attributes that include a) compactness, minimal footprint and size, b) negligible heat and mass transfer resistances, c) modularity, hence also load-following capability, d) shop fabrication, easy transportability and installation, e) minimum catalyst need and easy regenerability, and f) distributed applications. DIFREX provides services in building energy distributed networks for liquid fuel generation and supply, using microchannel reactors and renewable resources.

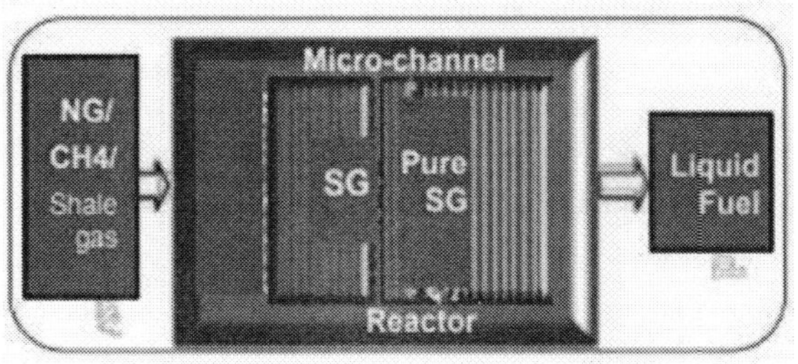

Figure 19: Micro-channel reactors for GTL applications

Value Chain Reactors: Fuels and chemicals from Syngas

Value chain reactors (**Figure 20**) can be easily and promptly developed and commercialized.

PTF (Power to Fuel) Technology

The cost of solar power, the most sustainable power source, is dropping and is contributing strongly to the US and world economy. According to one report, in 2016 the American solar industry added jobs 12x faster than the rest of the economy. There is a worldwide drive, particularly in Europe, to capitalize on the cheaper solar power to produce fuels and chemicals. Another area of key interest is to convert both powerplant exhaust and atmospheric CO_2 to fuel, especially to *methanol* that is dubbed as the *fuel of the future*. An economic *PTM* (Solar *Power To Methanol*) technology for conversion of atmospheric CO_2 to methanol would accomplish *zero-carbon footprint* in power consumption and utilization. We should heed the call for developing this emerging technology.

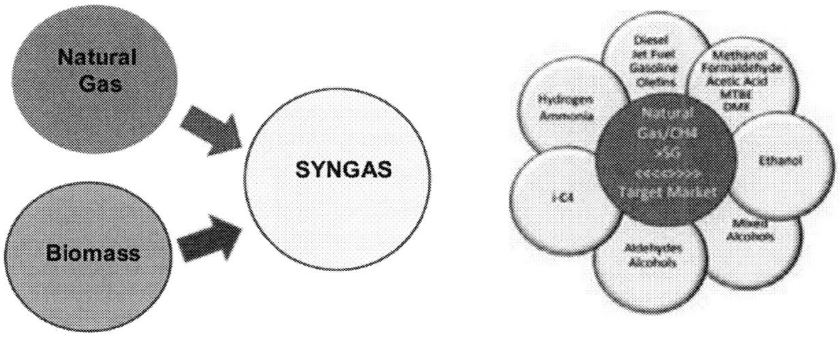

Figure 20: Conversion route of Value Chain Reactors

References:

1. Dutta, S and Gualy, R, "Build Robust Reactor Models", Chemical Engineering Progress, October, 2000, p. 37.
2. DIFREX[R] LLC NEXGEN Reactor Design - Fast assured reactor technology solutions for development, design, optimization, troubleshooting and revamp of commercial reactors. [cited November 13, 2017]; available from:www.difrex.com.
3. Dutta, S. and R. Gualy, "Super-modeling Reactors for Profitability", Chem. Eng'g. 107(6), 2000, p. 72.
4. Dutta, S. and R. Gualy, "Modernize Process Reactors", Hydrocarbon Processing, November, 1999, p. 91.
5. Dutta, S. and R. Gualy, "Overhaul Process Reactors", Hydrocarbon Processing, September, 1999, p. 43.
6. Dutta, S. and R. Gualy, "General Reactor Model Improves HPI Applications", Hydrocarbon Processing, July, 1999, p. 45.
7. Dutta, S., "Circulating Versus Bubbling Fluidized Bed for Hot Gas Desulfurization by Zinc Titanate", Proc. 5th Int'l. Circulating Fluidized Bed Conf., Beijing, PRC, May, 1996, p. pr-7.
8. Dutta, S., "Reactor Model: Usefulness of a Lesser Known Tool in Process Plants", paper presented at SABIC Tech. Meeting, May, 1996.
9. Dutta, S., "Fluidized and Moving Bed Desulfurization by Zinc Titanate", AIChE Symp. Ser. 301 (90), 1994, p. 157.
10. Dutta, S. and B. Jazayeri, "Alternate Reactor Concepts for Oxidative Coupling of Methane", Proc. 7th Int'l. Fluidization Conf., ed. Potter and Nicklin, Brisbane, Australia, May, 1992, p. 445.
11. Dutta, S. and G. D. Suciu., "An Experimental Investigation on the Effectiveness of Baffles in Breaking Bubbles in Fluid Beds", J. Chem. Eng'g. Japan 25 (3), 1992, p. 345.
12. Dutta, S., "A Novel CFB Application: A Chemical Compressor", Proc. 3rd Int'l. Circulating Fluidized Bed Conf., ed. Basu, Horio and Hasatani, Nagoya, Japan, Oct., 1990, p. 465.
13. Dutta, S., "A Novel Circulating Bed Metal Hydride Compressor", paper presented at the 8th World Hydrogen Conf., Honolulu, Hawaii, 1989.
14. Dutta, S., B. R. Christian and M. F. Ratterman, "Circulating Fluid Bed Reactor Model Developed for FCC/ART Catalyst Regenerator", Proc. 6th Int'l. Fluidization Conf., ed. Grace, Shemilt and Bergougnou, Banff, Canada, May, 1989, p. 9.
15. Dutta, S. and G. D. Suciu, "Unified Model Applied to the Scale-up of Catalytic Fluid Bed Reactors of Commercial Importance", Proc. 6th Int'l. Fluidization Conf., ed. Grace, Shemilt and Bergougnou, Banff, Canada, May, 1989, p. 311.
16. Dutta, S., S. C. Arnold, G. D. Suciu and L. Verde, "Scale-Up of a Commercial Catalytic Fluid Bed Reactor Involving Complex Kinetics", Ind. Chem. Eng'g. Symp. Ser. 87, 1984, p. 517.
17. Dutta, S., "Measurement of Bubble Characteristics by Optical Probe", paper presented at the AIChE Annual Meeting, Chicago, Nov., 1980.
18. Dutta, S., "A Simple Probe for Fluidized Bed Measurements", Cana. J. Chem. Eng'g. 57, 1979, p. 115.
19. Dutta, S., "Dead-zone Heights near the Grids of Fluidized Beds", Fluidization, Cambridge University Press, 1978, p. 32.
20. Wen, C. Y. and S. Dutta, "Research Needs for Analysis, Design and Scale-up of Fluidized Bed Reactors", AIChE Symp. Ser. 73 (161), 1977, p. 1.
21. Fluidization Engineering, 2nd edition, by Daizo Kunii and Octave Levenspiel, Butterworth-Heinemann, 1991.
22. Chemical Reaction Engineering, 3rd edition, by Octave Levenspiel, John Wiley & Sons, 1999.
23. Werther, J., Chem. Eng'g. Sci. 35, 1980, p. 372.
24. Kato, K., and C. Y. Wen, Chem. Eng'g. Sci. 24, 1969, p. 1351.
25. Hatano, H., T. Okuma and M. Ishida, J. Chem. Eng'g. Japan 17, 1984, p. 491.
26. Kwauk, M. N., N. Wang, Y. Li, B. Chen and Z. Shen, Circulating Fluidized Bed Technology, ed. P. Basu, Pergamon Press, 1986, p. 33.
27. Yang, W. C. and D. L. Keairns, AIChE Symp. Ser. 141(70), 1974, p. 27.
28. Wen, C.Y. and S. Dutta, Book Chapter "Kinetics of Coal Pyrolysis and Gasification Reactions", Ch. 2, Coal Conversion Technology, Addison-Wesley Publishing Company, 1979.
29. Wen, C.Y. and S. Dutta, "Solid-Gas Reactions in Coal Conversion Processes", "Coal Processing Technology", CEP Tech. Manual IV, 1978, p. 40.
30. Wen, C.Y. and S. Dutta, "Reactivity of Coal and Char: Part I – In Carbon Dioxide Atmosphere", I&EC Process Design & Development 16, 1977, p.20.
31. Wen, C.Y. and S. Dutta, "Reactivity of Coal and Char: Part II – In Oxygen-Nitrogen Atmosphere", I&EC Process Design & Development 16, 1977, p.31.
32. www.difrex.com.
33. http://difrex.com/pdf/BrochureContents.pdf
34. Dutta, Subhash, Stephen Arnold, Aashish Gaurav and Jim Brenner, "DIFREX Reactor and Technology Solutions for Many Particle Types and Sizes in Catalytic and Non-Catalytic Reaction Systems", paper presented at WCPT8 (8th World Congress on Particle Technology), Orlando, April, 2018.

Rare but Conceivable: Determining the Likelihood of Meteors and Other Infrequent Events

Michael S. Schmidt
Engineering Management and Chemical Engineering Departments
Missouri University of Science and Technology (MS&T)
and
Bluefield Process Safety
3101 Cherokee St.
St. Louis, MO 63118
MSchmidt@BluefieldSafety.com

Maria Vega-Westhoff
Bluefield Process Safety

Prepared for Presentation at
American Institute of Chemical Engineers
2018 Spring Meeting and 14th Global Congress on Process Safety
Orlando, Florida
April 22 – 25, 2018

Rare but Conceivable: Determining the Likelihood of Meteors and Other Infrequent Events

Michael S. Schmidt
Engineering Management and Chemical Engineering Departments
Missouri University of Science and Technology (MS&T)
and
Bluefield Process Safety
3101 Cherokee St.
St. Louis, MO 63118
MSchmidt@BluefieldSafety.com

Maria Vega-Westhoff
Bluefield Process Safety

Keywords: External Event Analysis, Facility Siting Data, Risk Assessment

Abstract

In the aftermath of a rare but devastating disaster, there is a tendency to overestimate how frequent such an event may be, hence a tendency to overestimate the risk associated with such an event. Unlike "black swans," which are inconceivable until they occur, events like meteors, hurricanes, and airplane crashes that impact chemical plants are rare, but conceivable. They occur with calculable likelihood and have quantifiable consequences, but most would not fault a PHA or LOPA team for dismissing a meteor strike as "not credible".

This paper considers several "meteors", both natural and man-made, that often become a cause of general concern after such an event occurs: meteors, earthquakes, and tsunamis; tornadoes and hurricanes; and airplane crashes, pipeline ruptures, train derailments, and truck crashes. It describes approaches to estimating the frequency of such events at a specific facility, and provides specific methods given certain assumptions. With these frequencies in hand, the paper goes on to develop a framework for giving meaning to the term "credible".

1 Conceivable *vs.* Credible

Nassim Taleb's Black Swan Theory can explain or excuse all sorts of catastrophic events. When Taleb repurposed the term "black swan" in 2007 [1], he described it this way: "First, it is an *outlier*, as it lies outside the realm of regular expectations, because nothing in the past can convincingly point to its possibility. Second, it carries an extreme impact. Third, in spite of its outlier status, human nature makes us concoct explanations for its occurrence *after* the fact, making it explainable and predictable." In other words, it is not simply rare but inconceivable, which means that it is not credible by any definition of the word. The fact that an explanation can be constructed after the event, in hindsight, does not mean that it was in any way predictable.

As a result, the Black Swan Theory is an interesting construct, but has no practical place in risk assessments. By their very nature, black swans are not something we can assess prospectively. We can only study them retrospectively.

1.1 "Meteors"

Catastrophic meteor strikes are not black swans. While quite rare, rarity is not what qualifies an event as a black swan. Black swans are not impossible, but they are inconceivable. Meteor strikes are not inconceivable. There is ample evidence from the past to convincingly point to their possibility.

Most meteors entering the earth's atmosphere explode in the upper atmosphere, where the solids vaporize and pose no threat to human activity. Asteroids larger than 20 m across (about the size of a barn) can make it close enough to the earth's surface that either the airburst or the impact crater have an effect. The Chelybinsk meteor (Figure 1) that exploded over Russia in 2013 had an estimated diameter of approximately 20 m; the airburst was about 500 kTons, around 30 times greater than that of the Hiroshima bomb.

More recently, a bolide streaked over Michigan. The boom from the exploding meteor registered in Michigan as an earthquake at 2.0 on the Richter Scale. A still taken from the security cameras at a chemical process facility in southern Michigan (Figure 2) tags the event at 8:08 pm EST on January 16, 2018.

Clearly, destructive bolides in the vicinity of chemical facilities are credible.

As metaphors, "meteors" should be far more interesting to risk analysts than "black swans". "Black swans," are inconceivable and so cannot be evaluated for risk. Once they occur, they are no longer inconceivable and hence, no longer "black swans". At that point, they can become part of the spectrum of potential events evaluated by future risk analysts.

"Meteors," on the other hand, are conceivable before they occur. They are rare, but occur at frequency that can be determined. When they are sufficiently rare, risk analysts will assume the catastrophic events they trigger are "not credible".

Photo credit: http://www.pbs.org/wgbh/nova/earth/meteor-strike.html

Figure 1. The Chelybinsk meteor that exploded over Russia on February 15, 2013.

Figure 2. The meteor that exploded over southern Michigan on January 16, 2018.

1.2 Credible scenarios

To allay concerns that imaginations will run wild, most analysts suggest limiting process hazard analyses and quantitative risk assessments to "credible scenarios". This then invites questions about what is and is not credible.

Exactly what does "credible" mean?

For some, credible means "possible" or "not impossible." This definition reduces "not credible" to "does not violate the laws of nature." Evil twins are no longer a soap opera plot device, but credible. Zombies, on the other hand, are still not credible.

For some, credible means something that has happened, as in "if it happened once, it could happen again". An aircraft once flew into a chemical plant in the southeast United States. As unlikely as that is to happen again, personnel at that plant deem any hazard review that does not consider being struck by an airplane as incomplete.

It is not the intent of this paper to declare what frequency is sufficiently rare that it can be treated as "negligible" or "not credible". Instead, it is the intent of this paper to give guidance in calculating the frequency of certain "meteors" so that risk analysts can make informed assessments of whether those events are sufficiently rare to treat as not credible, or conversely, determine the frequency to consider "not credible" based on the frequency of events that they do not consider credible.

2 External Causes — Improbable But Not Impossible

Consider nine improbable events that can not only result in catastrophic releases, but are of such a nature that analysts encounter them periodically while conducting risk assessments. The intent of this section is to address those events as potential causes for Layers of Protection Analysis (LOPA) and give guidance to analysts, so they can calculate frequency at which these events may impact their specific facility. This section divides the nine events into three categories: geophysical, climate-related, and transportation-related.

The geophysical events:

- Meteors
- Earthquakes
- Tsunamis

The climate-related events:

- Hurricanes
- Tornados

The transportation-related events:

- Pipeline ruptures
- Train derailments
- Truck crashes
- Airplane crashes

2.1 Geophysical

2.1.1 Meteors

An excellent and well annotated discussion of meteor impacts with earth [2] explains that while asteroids, comets, and meteoroids frequently impact the earth, the vast majority burn up or explode on entry into the earth's atmosphere as airbursts. We call these exploding meteors bolides and typically they have no effect on the surface of the earth. An asteroid must be at least 20 m across before the explosion is strong enough and at a low enough altitude to cause damage or before it can strike the earth and form a crater.

Using a table constructed with an on-line tool developed at Imperial College London [3], we determined that asteroids 20 m across or bigger impact the earth about once every 40 years. Typically, a once-every-40-year event will get the attention of a HazOp team. However, these events usually do not affect the entire world—the affected area is more like 250 km^2 (100 mi^2). The surface of the earth is about $5.2 \cdot 10^8$ km^2 (200 million mi^2), meaning that the likelihood that an asteroid strike will have an impact on a specific plant is low:

$$1 \text{ event/40 years} \cdot 250 \text{ km}^2 / 5.2 \cdot 10^8 \text{ km}^2 = 1.2 \cdot 10^{-8} \text{ events per year.}$$

2.1.2 Earthquakes

The occurrence of earthquakes around the globe follows a straight-forward relationship: as the earthquakes get 10 times stronger, they occur 10 times less often [4, 5]. Table 1 shows the rate of occurrence, as well as the effects of earthquakes by category.

Table 1. Earthquake Frequency

Category	**Moment Magnitude (Mw)**	**Effect**	**Global Occurrences**
Great	≥ 8	Causes serious damage in areas hundreds of miles across	1.3/yr
Major	7 – 7.9	Causes serious damage over larger areas	13/yr
Strong	6 – 6.9	Destructive in areas up to 160 km (100 mi) across	130/ yr
Moderate	5 – 5.9	Causes major damage to poorly constructed building over small regions; slight damage to well-design buildings	1,300/yr
Light	4 – 4.9	Noticeable shaking of indoor items; rattling noises	13,000/yr
Minor	3 – 3.9	Often felt	130,000/yr
Micro	2 – 2.9	Generally not felt but can be detected with instruments	1,300,000/yr

What should be clear from Table 1 is that a facility risk assessment need only consider the top three categories—Great, Major, and Strong. Worldwide, there are less than 150 earthquakes of interest per year. Unlike meteors, however, earthquakes do not occur randomly across the planet. Popular culture has made us all acutely aware of the "Ring of Fire" around the Pacific, as well as other seismic zones. The approach taken with meteors, that they are purely random events with equal likelihood anywhere on the planet, is not appropriate.

Instead, a map showing the frequency of earthquakes is the appropriate approach. Or more correctly, a map showing the frequency of damaging earthquake effects is the appropriate approach. Fortunately, at least for facilities in the United States, the U.S. Geological Survey [6] has prepared such a map (Figure 3). The map shows zones going from "cool," meaning less than 2 destructive earthquakes per 10,000 years, to "hot," which has more than 250 destructive earthquakes per 10,000 years. The upper Midwestern United States and the Gulf Coast are in the coolest earthquake zone. Facilities there can expect $1 \cdot 10^{-4}$ damaging earthquakes per year. Coastal California, central Alaska and the Aleutian Islands, and the northwest corner of Puerto Rico, on the other hand, are in the hottest earthquake zone and facilities there can expect $5 \cdot 10^{-2}$ damaging earthquakes per year.

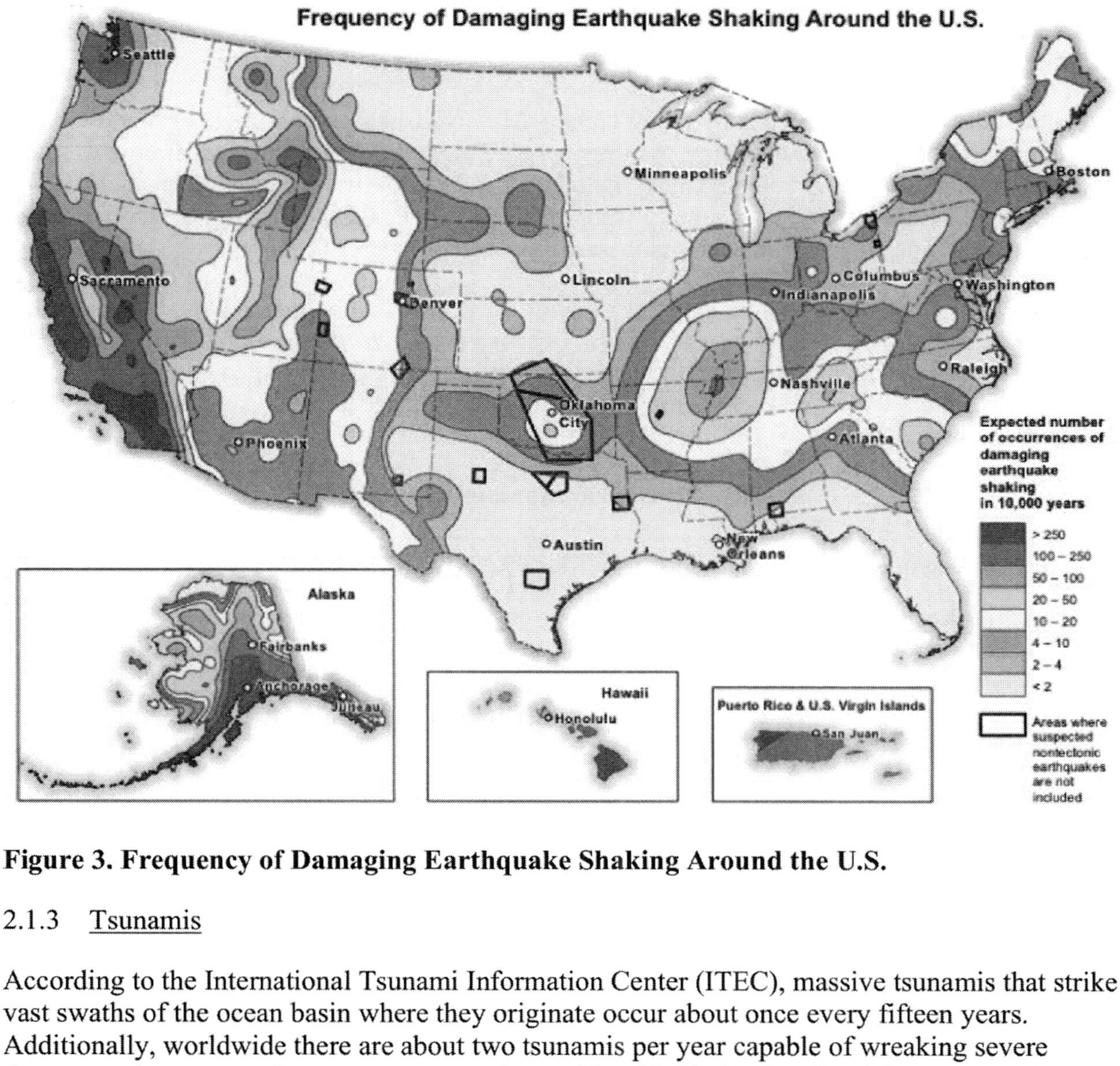

Figure 3. Frequency of Damaging Earthquake Shaking Around the U.S.

2.1.3 Tsunamis

According to the International Tsunami Information Center (ITEC), massive tsunamis that strike vast swaths of the ocean basin where they originate occur about once every fifteen years. Additionally, worldwide there are about two tsunamis per year capable of wreaking severe damage on large areas. Less severe tsunamis capable of inflicting localized damage occur even more often. While tsunamis can and have occurred in almost every large body of water, the Pacific Ocean basin and its associated waters are the most vulnerable [7].

In addition to the frequency of tsunamis, the associated impact zone is of interest to risk analysts. The impact zone is dependent on the size of the tsunami, which is dependent on the energy released by the initiating event—typically an earthquake, but landslides and volcanoes also trigger tsunamis—and the breadth of the initiating event. Tsunamis originate not just at the epicenter of an earthquake, but all along its length. Most data on the size of tsunamis focus on the height of the waves. However, for risk assessment, the breadth of the tsunami and inland incursion are of greater interest.

Inland incursion is a function of the height of the tsunami and the shore topography. Flat shoreline is more vulnerable; for instance, anything less than 30 m (100 ft) above sea level is vulnerable to a 30 m wave. Most tsunamis have a height of 3m (10 ft) or less. As a result, inland flooding as far as 300 m (1000 ft) is not unexpected, although some tsunamis have caused flooding 10 times further inland, with waves 10 times higher [8].

The breadth of tsunamis depends on many factors. Massive tsunamis can impact several hundred to several thousand miles of coastline. As they become bigger, they become less frequent, so that the risk of tsunamis large enough to cause any damage is essentially constant, independent of size. Collectively, it is reasonable to assume 5 tsunamis per year capable of causing damage to 20 km (12 mi) of coastline. This is consistent with data reported by the ITEC [9].

Table 2 shows the distribution of tsunami by body of water. Table 2 also shows the amount of coastline associated with each body of water. (Because of the fractal problem, there is generally little consensus on the length of coastline associated with an irregularly shaped body, whereas area can be calculated with great precision. Therefore, this article estimates coastline using the relationship of $10.5(\text{Area})^{0.5} = \text{Perimeter}$.)

Table 2. Distribution of Tsunamis

Body of Water	**Percentage of Tsunamis**	**Coastline Length**
Pacific Ocean	71%	135,000 km (84,000 mi)
Mediterranean Sea	15%	17,000 km (10,000 mi)
Atlantic Ocean and Caribbean Sea	7%	110,000 km (68,000 mi)
Indian Ocean	6%	89,000 km (55,000 mi)
Black Sea	1%	7,000 km (4,000 mi)

Consider a plant on the Pacific coast in California, within 300 m (1000 ft) of the shore and at an elevation of less than 30 m (100 ft) above sea level. The frequency at which a damaging tsunami originating in the Pacific Ocean would strike the 20 km stretch of coastline on which the plant operates is:

$$5 \text{ events/yr} \cdot 0.71 \cdot 20 \text{ km} / 135{,}000 \text{ km} = 0.0005 \text{ events/year} = 5 \cdot 10^{-4} \text{ events per year}$$

A similar plant situated on the Atlantic coast in Georgia would be at less risk:

$$5 \text{ events/yr} \cdot 0.07 \cdot 20 \text{ km} / 110{,}000 \text{ km} = 0.00006 \text{ events/year} = 6 \cdot 10^{-5} \text{ events per year}$$

Either plant significantly reduces the hazard by being more than 100 m (330 ft) above sea level, however far back from the shore that must be.

A Note About the Fukushima Nuclear Facility.

The Fukushima Nuclear Facility in Japan sat on coastline in the Pacific Basin. Before the catastrophic tsunami on March 11, 2011, the frequency at which a damaging tsunami would have struck the facility could be estimated at $5 \cdot 10^{-4}$ events per year. In hindsight, pundits and the popular media were quick to say, "they should have known." That destructive tsunamis are possible is not in dispute. The fact that one happened in 2011 is proof of that. The fact that one happened in 2011 is not proof that they happen frequently. After the 2011 tsunami, the frequency at which a damaging tsunami would have struck the facility is unchanged: about $5 \cdot 10^{-4}$ events per year.

2.2 Weather-Related

2.2.1 Hurricanes

The damage by hurricanes is from wind and storm surge. Storm surge is too dependent on the topography of the specific location and other factors for this paper to address, but a reasonable range is 1 to 10 m (3 to 30 ft) of elevation above sea level [10]. The impact zone of hurricane winds is a function of how far inland it travels before winds weaken and the span of the hurricane when it makes landfall. Typically, meteorologists mention 160 km (100 mi) as the distance a hurricane will travel inland before weakening to a tropical storm. The span of hurricane-force winds varies from 50 to 500 km (30 to 300 mi), with an average diameter of hurricane-force winds of 160 km (100 mi) [11].

On average, 1.77 hurricanes per year struck the United States during the years from 1851 to 2010 [12]. Meteorologists categorize hurricanes on the Safir-Simpson scale. While the frequency of hurricanes decreases with increasing strength, the Safir-Simpson scale, unlike the earthquake magnitude scales, is not logarithmic, so there is not an obvious distribution, as Table 3 shows.

Table 3. Hurricane Strength

Safir-Simpson Category	Description	Distribution by Category
1	119 – 153 km/h (74 – 95 mph)	40%
2	154 – 177 km/h (96 – 110 mph)	26%
3	178 – 208 km/h (111 – 129 mph)	26%
4	209 – 251 km/h (130 – 156 mph)	6%
5	> 252 km/h (> 156 mph)	1%

While the Safir-Simpson scale categorizes hurricanes, it is important to keep in mind that the scale is based solely on peak wind speed. The size and destructive energy of a storm and the intensity of the associated storm surge is only loosely correlated to a storm's Safir-Simpson category.

More relevant to a risk analyst considering a hurricane as an initiating event is the frequency at which it will strike a specific location. As was the case with tsunamis, hurricanes can strike any seacoast, but some locations are far more vulnerable than others. In the United States, only the coasts of the Gulf of Mexico and the Atlantic Ocean are vulnerable. Data is available for the eighteen coastal states bordered by these two bodies of water [12], and after normalizing for the length of coastline of each state, it is clear to see that some states are more vulnerable to hurricanes than others, as Table 4 shows, beginning at the southwest coastline in Texas and proceeding east and then north to Maine.

Table 4. Hurricane Frequency

State	Hurricanes per Century per 160 km (100 mi) of Coastline
Texas	8
Louisiana	8
Mississippi	16
Alabama	16
Florida	8
Georgia	8
South Carolina	8
North Carolina	8
Virginia	4
Maryland	4
Delaware	4
New Jersey	4
New York	4
Connecticut	4
Rhode Island	4
Massachusetts	4
New Hampshire	4
Maine	2

The estimates of coastline length have the same fractal problem described earlier. In this instance, the coastline length was based on a resolution of 30 minutes of latitude, or about 56.6 km (34.5 mi). Expressing the frequency per 160 km (100 mi) of coastline puts the values in terms of average hurricane diameters so that the values expressed give the frequency at which any location on these coasts can expect a hurricane to strike, per century.

Once the hurricane frequency for a particular location is determined, the facility should be assumed to be vulnerable to storm surge when is stands at an elevation less than 10 m (30 ft) and vulnerable to hurricane-force winds if it is within 160 km (100 mi) of the coast.

A Note About the 2017 Hurricane Season.

Before it was even over, the popular press had dubbed 2017 as The Year of the Hurricane. Three of the five costliest hurricanes on record, adjusted for inflation, occurred in 2017: Harvey (2nd, at $125 billion), Maria (3rd, at $90 billion), and

Irma (5th, at $50 billion). This compares to Katrina in 2005 (1st, at $160 billion) and Sandy in 2012 (4th, at $70 billion) [13]. That the costliest hurricanes are all recent, however, is not a reflection of a new power in hurricanes, but of extent of development on the coastlines, hence the amount of damage a hurricane can do. When a hurricane makes landfall on an undeveloped coastline, it does no damage at all, no matter how powerful it is.

Researchers have noted that there has been a recent uptick in hurricane activity in the Atlantic. It may be the beginning of an upward trend, meaning that past data is no longer relevant, or it may be normal statistical variation. At most, it is too early to tell [14]. After the chemical fires at the Arkema plant in Crosby, Texas, following the unprecedented flooding brought on by Hurricane Harvey in August 2017, some suggested that relying on existing data about hurricanes and flooding was no longer valid [15, 16]. In the aftermath of a rare but devastating disaster, there is a tendency to overestimate how frequent such an event may be. Hence, there is a tendency to overestimate the risk associated with such an event, to declare, "This is the new normal." However, in the absence of new data to replace the existing data, the existing data remains the best data there is. Anything else is pure speculation.

2.2.2 Tornados

Tornadoes are another violent weather-related event that are sometimes a concern. Like earthquakes, tsunamis, and hurricanes, the frequency of tornadoes is highly dependent on location. Most are aware of the region of North America known as "Tornado Alley" which extends northward from Texas through the plain states of Oklahoma, Kansas, Nebraska, South Dakota, North Dakota, and into the Canadian prairie provinces of Manitoba and Saskatchewan. However, nearly every state east of the Rocky Mountains experiences several tornadoes per year, as Figure 4 shows, as do Ontario, Manitoba, Alberta, and Saskatchewan.

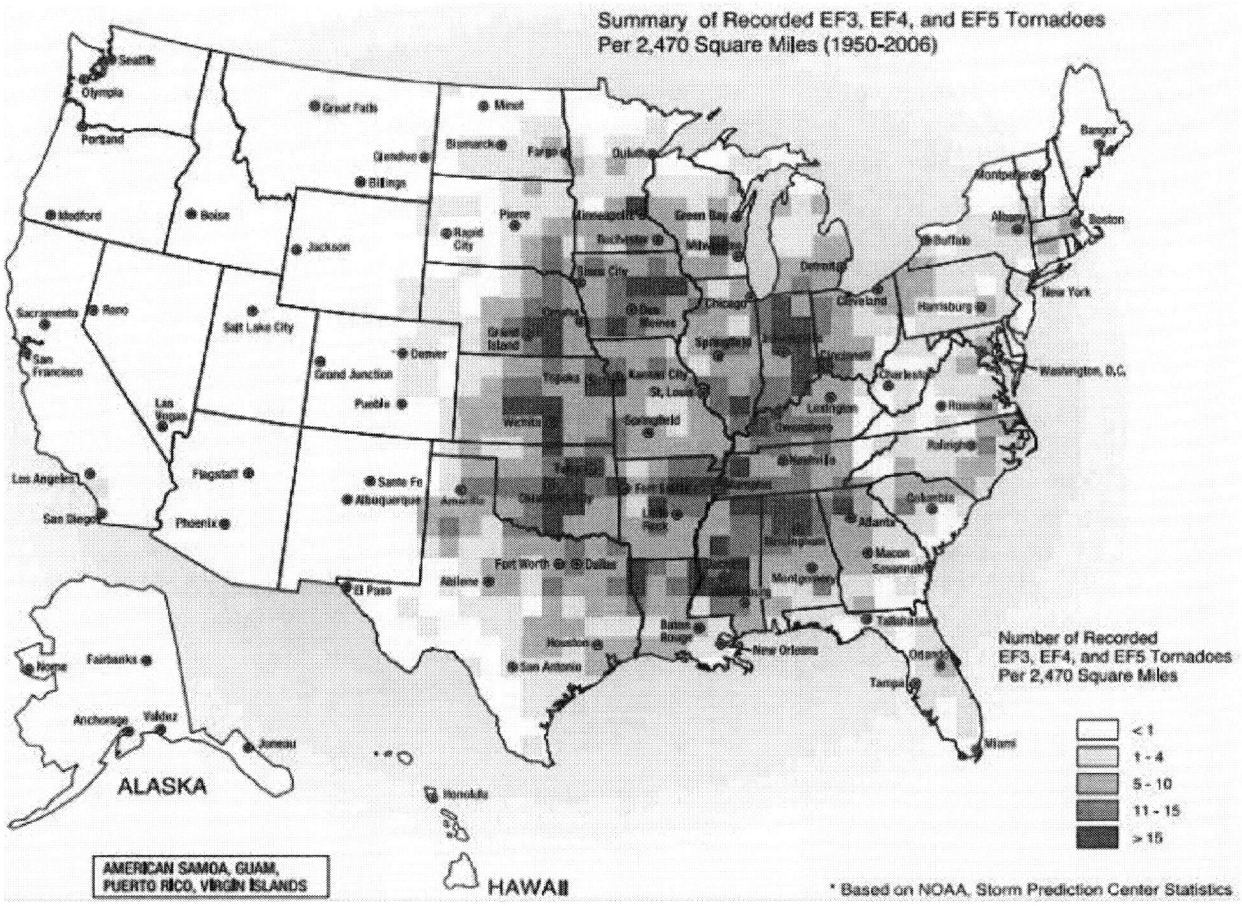

Figure 4. Total Tornado Activity in the United States – 1950 to 2006.

The impact zone of a tornado depends on the width and length of its path. The width of tornado paths ranges from 100 m (300 ft) to 4.2 km (2.6 mi); the typical width is about 1.6 km (1 mi). The longest observed path of a tornado is 352 km (215 mi), but tornado paths are rarely longer than 25 km (15 mi). The average travel speed of tornados is 50 km/h (30 mph). Because they typically last around 10 minutes, the average path of a tornado is 8.3 km (5 mi) [17]. The average impact zone of a tornado, then, is 13 km^2 (5 mi^2).

The blocks in Figure 4 use a color code to show the number of EF3, EF4, and EF5 tornadoes per 2,470 mi^2 per the 57 years covered by the analysis [18]. In other words, tornadoes per $1.4 \cdot 10^5$ mi^2 per year. Expressing the frequency in terms of impact zones is more relevant, however, so the most appropriate units are tornadoes per impact zone per year. The conversion appears in Table 5.

Table 5. Tornado Rates for Figure 4

Legend	NOAA	Frequency at which a tornado is likely to strike a facility
Very Low ☐	< 1 tornadoes per 2470 mi^2 per 57 yr	< $0.4 \cdot 10^{-4}$ tornadoes per 13 km^2 (5 mi^2) per yr
Low	1-4 tornadoes per 2470 mi^2 per 57 yr	0.4 - $1.5 \cdot 10^{-4}$ tornadoes per 13 km^2 (5 mi^2) per yr
Medium	5-10 tornadoes per 2470 mi^2 per 57 yr	1.8 - $3.5 \cdot 10^{-4}$ tornadoes per 13 km^2 (5 mi^2) per yr
High	11-15 tornadoes per 2470 mi^2 per 57 yr	4 - $5 \cdot 10^{-4}$ tornadoes per 13 km^2 (5 mi^2) per yr
Very High	> 15 tornadoes per 2470 mi^2 per 57 yr	> $5 \cdot 10^{-4}$ tornadoes per 13 km^2 (5 mi^2) per yr

Figure 4 is very granular, but it only includes the most extreme tornadoes with Enhanced Fujita scale ratings of EF3, EF4, and EF5. Using a state-by-state compilation [19] of confirmed tornadoes of all ratings, EF0 to EF5, Table 6 shows the frequency at with a tornado is likely to strike a facility, again normalizing the tornado frequency to a typical impact area of 13 km^2 (5 mi^2). With tornado frequencies for each state [19] and the area of each state (expressed as tornado impact zones), it is possible to determine tornado rates, as shown in Table 6. The rates shown in Table 6 differ from those shown in Figure 4, in that the Figure 4 rates only include EF3, EF4, and EF5 tornadoes, while the rates in Table 6 include all tornadoes, EF0 to EF5.

Table 6. Tornado Rates

State	**Frequency at which a tornado is likely to strike a facility** [tornadoes per 13 km^2 (5 mi^2) impact zone per year]
Kansas	$5.6 \cdot 10^{-3}$
Oklahoma	$4.7 \cdot 10^{-3}$
Illinois	$4.7 \cdot 10^{-3}$
Mississippi	$4.7 \cdot 10^{-3}$
Alabama	$4.5 \cdot 10^{-3}$
Iowa	$4.4 \cdot 10^{-3}$
Florida	$4.2 \cdot 10^{-3}$
Maryland	$4.0 \cdot 10^{-3}$
South Carolina	$3.6 \cdot 10^{-3}$
Arkansas	$3.6 \cdot 10^{-3}$
Louisiana	$3.6 \cdot 10^{-3}$
Nebraska	$3.5 \cdot 10^{-3}$
Tennessee	$3.5 \cdot 10^{-3}$
Indiana	$3.4 \cdot 10^{-3}$
Missouri	$3.3 \cdot 10^{-3}$
Kentucky	$3.0 \cdot 10^{-3}$
Texas	$2.7 \cdot 10^{-3}$
North Carolina	$2.7 \cdot 10^{-3}$
Georgia	$2.5 \cdot 10^{-3}$
Minnesota	$2.4 \cdot 10^{-3}$
Colorado	$2.4 \cdot 10^{-3}$
North Dakota	$2.2 \cdot 10^{-3}$
Ohio	$2.1 \cdot 10^{-3}$
South Dakota	$2.1 \cdot 10^{-3}$
Virginia	$2.1 \cdot 10^{-3}$

State	**Frequency at which a tornado is likely to strike a facility** [tornadoes per 13 km^2 (5 mi^2) impact zone per year]
Delaware	$2.0 \cdot 10^{-3}$
Wisconsin	$1.8 \cdot 10^{-3}$
Pennsylvania	$1.7 \cdot 10^{-3}$
Connecticut	$1.4 \cdot 10^{-3}$
New Jersey	$1.1 \cdot 10^{-3}$
New York	$0.9 \cdot 10^{-3}$
Michigan	$0.8 \cdot 10^{-3}$
Massachusetts	$0.7 \cdot 10^{-3}$
Rhode Island	$0.6 \cdot 10^{-3}$
Wyoming	$0.6 \cdot 10^{-3}$
West Virginia	$0.5 \cdot 10^{-3}$

The frequency of tornadoes in the states west of the Great Plains and the states in northern New England is at a background level of $< 4 \cdot 10^{-4}$. This is 10 times higher than the "low" level shown in Figure 4, the difference being that Figure 4 only included EF3 to EF5 tornadoes, which account for less than 10% of all tornadoes, while Table 6 includes all tornadoes, EF0 to EF5. As for the 36 states on the list, the tornado leader is Kansas, which is what most would expect.

2.3 Transportation-Related

2.3.1 Pipeline ruptures

Pipeline ruptures result in approximately 25 barrels spilled per billion barrel-miles transported and 5.2 fatalities per trillion barrel-miles transported [20]. Risk analysts, however, are more concerned with the incident rate than spilled volume rate. The Center for Chemical Process Safety (CCPS) has addressed this specifically in its *Guidelines for Initiating Events and Independent Protection Layers in Layer of Protection Analysis* [21]. The committee determined that it would be useful to make two distinctions: between pipelines that are larger and smaller than 150 mm (6 inches) and between full breaches, sometimes called "guillotine" breaches, and partial breaches, or leaks, that amount to 10% of a full breach. The data in Table 7 refers to above-ground pipelines.

Table 7. Pipeline Failure Rates

Piping Size	Piping Failure Type	Frequency
Greater than 150 mm diameter (> 6 inches dia.)	Full breach (Guillotine failure)	$1 \cdot 10^{-5}$ failures per 100 m per year
Greater than 150 mm diameter (> 6 inches dia.)	Partial breach – 10% (Leak)	$1 \cdot 10^{-4}$ failures per 100 m per year
Less than or equal to 150 mm diameter (≤ 6 inches dia.)	Full breach (Guillotine failure)	$1 \cdot 10^{-4}$ failures per 100 m per year
Less than or equal to 150 mm diameter (≤ 6 inches dia.)	Partial breach – 10% (Leak)	$1 \cdot 10^{-3}$ failures per 100 m per year

The CCPS committee determined that there were too many factors to take into account to offer a generic failure rate for underground piping, although in general, underground piping failures are more frequent than above-ground piping failures, in large part because above ground pipelines are considerably easier to inspect.

2.3.2 Train derailments

Most discussions of train derailments report them in rates per train miles or similar metrics. Baldas reported that railroad crude oil spills were at the rate of 431 barrels per billion barrel-miles transported and that fatalities were at the rate of 15 fatalities per trillion barrel-miles transported [20].

Train derailments are quite frequent, but the vast majority occur at speeds of less than 5 mph with little or no damage. This is the sort of derailment that typically occurs within a plant fence line and while it can pose significant operational difficulties, there is generally no reason for this sort of incident to be the concern of a risk assessment team. Train derailments at speeds high enough to result in significant harm, such as the Lac-Mégantic rail disaster of July 6, 2013 that killed 47, are far less frequent. The "List of American railroad accidents" [22] notes 26 major American railroad accidents so far during the 21st century. However, somewhere between derailments that are daily inconveniences for operating facilities and major catastrophes is the number that risk analysts should consider.

The Federal Railroad Administration (FRA) reported that there were almost 19,000 train accidents in the ten-year period from 2008 through 2017. Of those, 71% were derailments [23]. Elsewhere, the FRA reported that there are 140,000 miles of railroad operated in the United States [24]. The rate of derailment, then, is 0.01 derailments per year per mile of railroad, or more relevantly, $6 \cdot 10^{-4}$ derailments per year per 100 meters of railroad.

2.3.3 Truck crashes

On a volume basis, tank truck crashes result in about 660 barrels of crude oil spilled per billion barrel-miles transported. Likewise, tank truck crashes result in about 80 fatalities per trillion barrel-miles transported. Both rates are worse than for transport by train cars and much worse than for transport by pipeline [20].

The incident rate for truck crashes are not reported on the basis of infrastructure—in crashes per mile of roadway—but on the basis of traffic, that is, crashes per 100 million vehicle miles travelled. Counting only those crashes that result in fatalities as sufficiently serious to warrant attention in a facility siting study, the average rate for the period from 2007 to 2015 is 1.24 fatal truck crashes per 100 million miles travelled by large trucks [25].

To determine the incident rate on the basis of crashes per mile of roadway requires that the analyst determine the length of roadway considered and then estimate the volume of truck traffic on that roadway. While there are portions of the National Highway System that carry more than 8,500 trucks per day [26], the vast majority of roadways carry considerably less. As an upper boundary, the incident rate for truck crashes is 0.04 crashes per year per mile of high-volume highway, or $2 \cdot 10^{-3}$ truck crashes per year per 100 meters of highway. For the roadways more likely to be of interest in a risk assessment, the volume is likely to be 10 to 100 times less. However, some sampling is in order before estimating the truck crash rate for a specific roadway.

2.3.4 Airplane crashes

As most would imagine, airplane crashes into process plants are exceedingly rare. A search of the literature only identified three in the United States—one in 1949 in Dallas, Texas [27], one in 1985 in Kansas City, Kansas [28], and one in 2016 in Lubbock, Texas [29]. There are undoubtedly others, but they are so rare that they should always be newsworthy; it is hard to imagine that many go unreported. A recent article discussed the risk of chemical plants near airports [30], but focused more on the consequences of those incidents than on a methodology for estimating the frequency of airplane crashes into chemical plants.

Berg has described a useful methodology for estimating the likelihood of an airplane crashing into a specific facility [31]. Beyond 10 km away from an airport, the likelihood is a function of the area of the facility, the distance from the airport of the facility, and the rate at which aircraft land at and take off from the airport. Within 10 km of an airport, the likelihood remains a function of these factors, but also increases exponentially as the distance decreases.

Berg's method can also account for the weight class of the airplane, the deviation from the landing/take-off axis, additional airports, and whether the aircraft are commercial, military, or other. These distinctions are only necessary, however, when the basic methodology suggests that the frequency is too high and would benefit from a more detailed analysis.

The fundamental relationship is:

$$H = h \cdot f \cdot A_{FACILITY} / (2\pi R \cdot d_{AVG})$$

Where

H	is the frequency of airplane crashes into the facility (crashes/year)
h	is the global crash rate
f	is the frequency of flights from the airport considered (flights/year)
$A_{FACILITY}$	is the area of concern for the facility (L^2)
R	is the distance of the facility from the airport considered (L)
d_{AVG}	is the average flight distance to and from the airport considered (L)

Note: L can be in any units, but it must be in the same units in all three instances.

The global crash rate, h, is $4 \cdot 10^{-7}$ crashes/flight. This is derived from a recent report from the International Air Transport Association (IATA) [32] which indicated a five-year average of 86 aviation accidents per year, of which ~20% were severe enough to result in the loss of the aircraft, or "hull loss" accidents. Hull loss accidents are the kind of loss that would affect a chemical plant [32]. For a facility occupying 250 acres (1.0 km^2) at a distance of 50 km (31 mi) from Saint Louis Lambert International Airport (STL) in Missouri, where there are 260 flight departures a day (95,000 flights/year) at an average flight distance of 1,200 km (750 mi), the annual frequency of airplane crashes into the facility would be $1 \cdot 10^{-7}$ crashes/year.

Facilities within 10 km of an airport experience an exponential climb in the crash rate the closer they are to the airport. Berg's work addresses this as well. The relationship, based on flight operations—take-offs and landings—rather than flights, is:

$$H = h \cdot e^{(-0.5/\text{km} \cdot R)} \cdot f \cdot A_{FACILITY} / (2\pi \text{ km} \cdot R)$$

Where

H	is the frequency of airplane crashes into the facility (crashes/year)
h	is the global landing/take-off crash factor ($1.3 \cdot 10^{-7}$ crashes/flight)
R	is the distance of the facility from the airport considered (km)
f	is the frequency of flights from the airport considered (flights/year)
$A_{FACILITY}$	is the area of concern for the facility (km^2)
2π km $\cdot$ R	is the area of the 1 km wide annular segment at distance R from the airport considered (km^2)

Considering an identical process facility, but only 5 km (3.1 miles) from STL, the annual frequency of airplane crashes into the facility would be $3 \cdot 10^{-5}$ crashes/year.

Strictly speaking, the relationship above only applies to facilities that are within 45° of the flight path for landing and take-off. A conservative assessment would treat all facilities as though they are within 45° of a flight path, particularly since most large airports have more than one runway.

3 Conclusions and Recommendations

Risk has two components, consequence and likelihood. It follows, then, that an analysis that only addresses consequences does not address risk. In the absence of likelihood, a consequence is not particularly meaningful. Decisions should be driven by risk assessments, not consequence assessments. Otherwise, every plant in the world should have an anti-meteor task force. Unlike "black swans," which are inconceivable until they happen, "meteors" are rare, but conceivable. However, most people would not know how to quantify how rare these "meteors" are.

When we considered nine types of conceptually rare events in terms of the likelihood of them impacting a process facility, it was clear that they were not only conceivable, but calculable. Except in the case of actual meteors, the frequency is a function of a facility's location and in some cases, other factors. Rather than use the numbers in Table 8, which compiles the examples used throughout this article (rounding to the nearest order of magnitude to make comparison easier), analysts are advised to do the simple math necessary to determine the likelihood for their specific facility.

Table 8. Example Event Frequencies

Facility Event Description	**Frequency (nearest order of magnitude)**
> 20 m meteor strike	$1 \cdot 10^{-8}$ per year
Damaging earthquake in upper Midwest	$1 \cdot 10^{-4}$ per year
Damaging earthquake on California coast	$1 \cdot 10^{-2}$ per year
Damaging tsunami on Pacific coast	$1 \cdot 10^{-3}$ per year
Damaging tsunami on Atlantic coast	$1 \cdot 10^{-4}$ per year
Hurricane on Gulf of Mexico coast	$1 \cdot 10^{-1}$ per year
Hurricane on New England coast	$1 \cdot 10^{-2}$ per year
Tornado in Tornado Alley	$1 \cdot 10^{-2}$ per year
Tornado in Rocky Mountains	$1 \cdot 10^{-4}$ per year
Rupture of large pipeline (> 150 mm, > 6 inches)	$1 \cdot 10^{-5}$ failures per 100 m per year
Significant train derailment	$1 \cdot 10^{-3}$ failures per 100 m per year
Significant truck crash on high volume highway	$1 \cdot 10^{-3}$ failures per 100 m per year
Significant truck crash on medium volume highway	$1 \cdot 10^{-4}$ failures per 100 m per year
Significant truck crash on low volume highway	$1 \cdot 10^{-5}$ failures per 100 m per year
Plane crash on plant 50 km from medium-sized airport	$1 \cdot 10^{-7}$ per year
Plane crash on plant 5 km from medium-sized airport	$1 \cdot 10^{-4}$ per year

When U.S. Supreme Court Justice Stevens wrote the majority opinion in the 1980 OSHA benzene case, he said, "Some risks are plainly acceptable and others are plainly unacceptable," and then went on to add that the odds of fatality of one in a billion could not be considered significant but that for odds of a fatality of one in a thousand "a reasonable person might well consider the risk significant" [33]. Given a range of one in a thousand years as plainly unacceptable and one in a billion years as plainly acceptable, none of these "meteors" meet the criteria of "plainly acceptable," not even meteors. On the other hand, there are several that meet the criteria of "a reasonable person's significant risk."

It is in that gap between once every thousand years and once every billion years that we will find the definition of "credible". Analysts should consider the frequencies of the events discussed here, and define "credible" for their organization. With that definition in hand, they will be able eliminate scenarios that are "not credible."

4 References

[1] Taleb, Nassim Nicholas, *The Black Swan: The Impact of the Highly Improbable (First Edition)*. Random House, New York. 2007.

[2] "Impact Event," *Wikipedia – The Free Encyclopedia*, The Wikimedia Foundation, San Francisco, CA, last edited 01-Feb-2018. Accessed 14-Feb-2018 at https://en.wikipedia.org/wiki/Impact_event#Frequency_and_risk .

[3] Marcus, R., H.J.Melosh, and G. Collins, "Earth Impact Effects Program: A Web-based computer program for calculating the regional environmental consequences of a meteoroid impact on Earth," *Meteoritics and Planetary Science*, 40:6, June 2005. pp. 817-840. Accessed 14-Feb-2018 at http://impact.ese.ic.ac.uk/ImpactEffects/

[4] Shearer, P.M. and P.B.Stark, "Global Risk of Big Earthquakes Has Not Recently Increased," Proceedings of the National Academy of Sciences of the United States of America, 109:3, November 12, 2011. pp. 717-721. Accessed 14-Feb-2018 at http://www.pnas.org/content/109/3/717.full .

[5] NEIC, "Earthquakes," *Earthquake Hazards Program*, National Earthquake Information Center, United States Geological Survey, U.S. Department of Commerce, Denver, CO. Accessed 14-Feb-2018 at https://earthquake.usgs.gov/earthquakes/ .

[6] NEIC, "Introduction to the National Seismic Hazard Maps," *Earthquake Hazards Program*, National Earthquake Information Center, United States Geological Survey, U.S. Department of Commerce, Denver, CO. Accessed 14-Feb-2018 at https://earthquake.usgs.gov/earthquakes/ .

[7] ITIC, "Where and how frequently are tsunamis generated?" *International Tsunami Information Center Web Site*, International Tsunami Information Center, Honolulu, HI. Accessed 14-Feb-2018 at http://itic.ioc-unesco.org/index.php?option=com_content&view=article&id=1163:where-and-how-frequently-are-tsunamis-generated&catid=1340&Itemid=2055 .

[8] National Geographic Staff, "Tsunamis: Facts About Killer Waves, National Geographic, 28-Dec-2004. Accessed 14-Feb-2018 at https://news.nationalgeographic.com/news/2004/12/tsunamis-facts-about-killer-waves/ .

[9] ITIC, "List of Tsunamis," *International Tsunami Information Center Web Site*, International Tsunami Information Center, Honolulu, HI. Accessed 14-Feb-2018 at http://itic.ioc-unesco.org/index.php?option=com_content&view=category&layout=blog&id=1160&Itemid=1077&lang=en .

[10] Masters, J., "Characteristics of Storm Surges, *Weather Underground*, The Weather Company, LLC, Atlanta, GA. Accessed 14-Feb-2018 at https://www.wunderground.com/hurricane/surge_characteristics.asp .

[11] Hasling, J.F., "What is the Size of a Hurricane?" *WRC Texas Hurricane Handbook*, Weather Research Center, Houston, TX. Accessed 14-Feb-2018 at http://www.wxresearch.org/family/pg5.html .

[12] Blake, E.S., C.W. Landsea, and E.J. Gibney, "The Deadliest, Costliest, and Most Intense United States Tropical Cyclones from 1851 to 2010," *NOAA Technical Memorandum NWS NHC-6*, National Hurricane Center, National Oceanic and Atmospheric Administration, Miami, FL. August 2011. 52 pp. Accessed 14-Feb-2018 at https://www.nhc.noaa.gov/dcmi.shtml .

[13] NHC Public Affairs, "Costliest U.S. Tropical Cyclones Tables Updated," Addendum to *NOAA Technical Memorandum NWS NHC-6*, National Hurricane Center, National

Oceanic and Atmospheric Administration, Miami, FL. January 26, 2018. 52 pp. Accessed 14-Feb-2018 at https://www.nhc.noaa.gov/news/UpdatedCostliest.pdf .

[14] Ramaswamy, V., "Global Warming and Hurricanes: An Overview of Current Research Results," *GFDL Web Page*, Geophysical Fluid Dynamics Laboratory, Princeton University, Princeton, NJ. 24-Jan-2018, Accessed 14-Feb-2018 at https://www.gfdl.noaa.gov/global-warming-and-hurricanes/ .

[15] Dempsey, M. and J. Carpenter, "Arkema documents: Planning, mechanical failures led to Harvey chemical fires," Houston Chronicle, 15-Nov-2017. Accessed 14-Feb-2017 at https://www.houstonchronicle.com/news/houston-texas/houston/article/Arkema-documents-show-planning-mechanical-12358188.php .

[16] Dunklin, R. "Agency: Texas chemical plant unprepared for Hurricane Harvey," Chicago Tribune, 17-Nov-2017. Accessed 14-Feb-2017 at http://www.chicagotribune.com/news/nationworld/sns-bc-us--harvey-chemical-plant-20171115-story.html .

[17] "About Tornadoes," *Weather Underground*, The Weather Company, LLC, Atlanta, GA. Accessed 01-Feb-2017 at https://www.wunderground.com/resources/education/tornadoFAQ.asp?MR=1 .

[18] FEMA, Taking Shelter From the Storm: Building a Safe Room for Your Home or Small Business, FEMA 320, 3rd Ed., Federal Emergency Management Agency, Department of Homeland Security, Washington, DC. August 2008. p. 3. Accessed 14-Feb-2018 at http://www.weather.gov/media/bis/FEMA_SafeRoom.pdf .

[19] Livingston, I., "Annual and monthly tornado averages for each state (maps), *U.S.Tornadoes*, 06-Apr-2016. Accessed 14-Feb-2018 at http://www.ustornadoes.com/2016/04/06/annual-and-monthly-tornado-averages-across-the-united-states/ .

[20] Baldas, A., "The Crude Truth: Assessing the Associated Risks of Rail, Tanker, and Pipeline Crude Oil Transport," *International Mechanical Engineering Congress and Exposition,* Safety Engineering and Risk Analysis Division, ASME. Montreal, Canada. November 2014.

[21] Chastain, W., et al., *Guidelines for Initiating Events and Independent Protection Layers in Layer of Protection Analysis*, Center for Chemical Process Safety, American Institute of Chemical Engineers, New York. 2015. pp. 105-112.

[22] "List of American railroad accidents," *Wikipedia – The Free Encyclopedia*, The Wikimedia Foundation, San Francisco, CA, last edited 04-Feb-2018. Accessed 14-Feb-2018 at https://en.wikipedia.org/wiki/List_of_American_railroad_accidents .

[23] FRA, "Ten Year Accident/Incident Overview," *FRA Office of Safety Analysis Web Site*, Federal Railroad Administration, U.S. Department of Transportation, Washington, DC. November 2017. Accessed 14-Feb-2018 at http://safetydata.fra.dot.gov/OfficeofSafety/publicsite/Query/TenYearAccidentIncidentOverview.aspx .

[24] FRA, "Freight Rail Overview," *FRA Web Site*, Federal Railroad Administration, U.S. Department of Transportation, Washington, DC. Accessed 14-Feb-2018 at https://www.fra.dot.gov/Page/P0362 .

[25] FMSCA, "Trends Table 4. Large Truck Fatal Crash Statistics, 1975-2015," *Large Truck and Bus Crash Facts*, Federal Motor Carrier Safety Administration, U.S. Department of Transportation, Washington, DC. February 2017. Accessed 14-Feb-2018 at https://www.fmcsa.dot.gov/safety/data-and-statistics/large-truck-and-bus-crash-facts-2015 .

[26] FHA, "Peak-Period Congestion on High-Volume Truck Portions of the National Highway System," Freight Analysis Framework, V4.3. Federal Highway Administration, U.S. Department of Transportation, Washington, DC. 2017. Accessed 14-Feb-2018 at https://ops.fhwa.dot.gov/freight/freight_analysis/nat_freight_stats/nhsconghvtrk2045.htm

[27] "American Airlines Flight 157," *Wikipedia – The Free Encyclopedia*, The Wikimedia Foundation, San Francisco, CA, last edited 19-Jun-2017. Accessed 14-Feb-2018 at https://en.wikipedia.org/wiki/American_Airlines_Flight_157 .

[28] McGuire, M.J. "January 9," *This Day in Water History*. Accessed 14-Feb-2018 at https://thisdayinwaterhistory.wordpress.com/2015/01/09/january-9-1985-plane-crashes-into-kansas-water-treatment-plant-2012-first-haitian-cholera-victim/

[29] AP, "Small plane that crashed into Texas cement plant was stolen from Midland," Lubbock Avalanche-Journal, Lubbockonline.com, 8-Jun-2016. Accessed 14-Feb-2018 at http://www.lubbockonline.com/filed-online/2016-06-08/small-plane-crashed-texas-cement-plant-was-stolen-midland .

[30] Konersmann, R. "Chemical Plants in the Vicinity of Airports—Assessing the Risk," *Chemical Engineering and Technology*, 32:2. 2009. pp.232-240.

[31] Berg, H.-P., "Risk Assessment of Aircraft Crash Onto a Nuclear Power Plant," *Reliability: Theory and Applications*, 01(20) (Vol. 2). March 2011. pp. 38-51.

[32] Gani, A., "How common are plane crashes?" *The Guardian*. 25-Jul-2014. Accessed 14-Feb-2018 at https://www.theguardian.com/world/2014/jul/24/how-common-are-plane-crashes .

[33] Stevens, J.P. writing for the majority, "Industrial Union Department AFL-CIO v. American Petroleum Institute et al." *Case No. 78-911*, U.S. Supreme Court, Washington, D.C. July 2, 1980. 448 U.S. 655.

Concept Risk Comparison of Transportation Means

Cynthia Spitzenberger
DNV GL
1400 Ravello Drive, Katy, TX 77449
cynthia.spitzenberger@dnvgl.com

Lindsay Deal
DNV GL

Prepared for Presentation at
American Institute of Chemical Engineers
2018 Spring Meeting and 14th Global Congress on Process Safety
Orlando, Florida
April 22 – 25, 2018

Concept Risk Comparison of Transportation Means

Cynthia Spitzenberger
DNV GL
1400 Ravello Drive, Katy, TX 77449
cynthia.spitzenberger@dnvgl.com

Lindsay Deal
DNV GL

Keywords: risk, transportation, QRA, highly volatile liquid

Abstract

Geographical distances between production and consumption of highly volatile liquids are increasing transport volumes across the continental United States. Supply chain planning requires comparison across transportation means. The complex process of selecting a transport means and route involves a range of parameters targeting both cost and feasibility. Risk to the public along the chosen route factors into both cost and feasibility comparisons.

This paper provides a framework to compare the risk to the public from a highly volatile liquid loss of containment event from pipeline, rail, truck and barge transportation. The risk-based framework normalizes the frequency on volume and distance transported and addresses consequences in terms of potential volume lost and population density to provide a direct, lateral comparison.

The presented risk results vary across transport modes and population densities, and are presented on an annual volume shipped basis. Given an annual volume amount and a specified population density, comparison of the different transport modes is possible on a coarse basis. However, route length likely varies between the different transport modes and would factor the expected risk value. Population density would also likely vary along the length of a transport route and would change the expected risk result.

Generally, for shipment of volumes less than 7 million gallons annually, the risk impact to public from the different transport modes in increasing order is barge, pipeline, rail and truck. For shipment volumes over 7 million gallons annually, pipeline transport presents the least risk exposure of the four transport modes.

1 Background

The transport of highly volatile liquids (HVL) within the US is a timely topic. The purpose of this evaluation is to provide a simplified and generic framework for the comparison of possible transport routes of HVL from a public risk perspective. To create a generic framework, a case study was performed to evaluate the HVL transport options. The postulated transport route selected is from the Commonwealth of Pennsylvania to a theoretical Gulf Coast refinery. Such a route allows comparison for the four transport options.

- Pipeline
- Rail
- Barge
- Truck

The case study considered specific transit locations for the paths of the different transport modes (existing pipeline and rail networks; rivers for barge transit; highways for hazardous transport). However, the details of the case study are not elaborated within the paper as they form only the basis to generalize the calculations to develop the framework. A normalized risk result is estimated for the comparison evaluation. This avoids calculation of societal risk metric which would be specific to location and frequency of operation. The key result aim is to draw conclusions from the generic framework regarding risk comparisons between the four modes of transit by volume transported. The generic result also provides a simplified comparison tool that could be used for coarse initial evaluations independent of the route.

The loading / unloading operations that would occur at either end of the route for the different transport modes are not considered. Cost evaluations for the different transport means are not included or considered.

2 Framework Basis

Parameters and assumptions that are common across transport modes include:

- Material
- Population categories
- Meteorological data
- Consequences of a release
- Limitations to the framework

DNV GL's Phast consequence tool was used for the consequence modeling.

2.1 Material

This study uses pure propane as a representative material for liquefied petroleum gas (LPG). However, highly volatile liquids can have a range of composition. For example, LPG is a mixture of materials highly dependent on its source.

2.2 *Population Categories*

Population densities drive both the public exposure to a consequence along the route and the probability of a delayed ignition.

Population densities are available from U.S. Census data by zip code allowing any route to be subdivided and quickly categorized with the application of geographic information system (GIS) software. This framework applies the Nuclear Regulatory Commission (NRC) definitions of Low, Medium and High population density categories (1) and develops continuous population density ranges from the 2010 U.S. Census data (2) to fit those definitions as shown in Table 1.

Table 1 Population Density Range

Population Density Category	NRC Representative Persons per square mile (1)	Population Density (persons per square mile) Range Applied
Low	15	0 – 47
Medium	1,862	48 – 6,142
High	10,000	6,143 + (max 156,667 (2))

2.3 *Meteorological Data*

The dispersion of a cloud of hazardous material is governed primarily by wind speed, wind direction and the atmospheric stability. Factors, which increase the dilution of a hazardous cloud with respect to distance traveled, are increasing wind speeds and decreasing stability of the atmosphere. However, high winds may transport hazardous materials far downwind before they become sufficiently diluted to no longer pose a hazard. An unstable atmosphere, typically experienced on a sunny day, causes increased vertical mixing, which further dilutes the hazardous clouds as they disperse downwind. The effect of wind direction is obvious in that only receptors downwind of the release are affected.

The meteorological data used in the Phast model consist of wind speed, humidity, solar radiation flux and ambient temperature.

Meteorological data vary along a given transport route in addition to the typical seasonal and weather pattern variations. Detailed analysis along a given route requires consideration of these variations. Initially, an average could be applied along the entire route with sensitivities used to bound how the risk results change at the extreme end of parameter ranges allowing a more detailed focus on the high-risk segments.

For simplicity, this framework applies a single combination of wind speed, humidity, solar radiation and ambient temperature since the goal is to compare similar routes across transport types. The weather condition applied is D stability with 5 m/s (11mph) wind speed, average atmospheric temperature of

50 degF. The maximum downwind distance of consequence effects is applied in all directions to account for variations in wind direction.

2.4 Consequences

Consequences are initially split between instantaneous (less than 20 seconds release duration) and continuous releases. Immediately ignited instantaneous releases result in a fireball. Immediately ignited continuous releases result in a jet or pool fire depending on the discharge rate.

All delayed ignition events are modeled as flash fires or vapor cloud explosions (VCE), where pool fires will accompany the flash fires/VCE for liquid spills. Most delayed ignition events also burn-back to form jet or pool fires that follow the initial flash fire or VCE, although the impacts to personnel are dominated by the initial flash fire/VCE effects.

Trucks, barges and railcars are assumed to stop upon an instantaneous release or upon ignition. However, small continuous releases may occur during transit. BLEVE (Boiling Liquid Expanding Vapor Explosion) hazards are considered as escalation events possible for the rail and truck transits.

Table 2 presents a summary of the different consequence hazards with the applied endpoint metrics and fatality fractions. Note that populations along the route are not assessed to be indoor or outdoor. Instead all population densities consider the same potential hazard exposure. The fatality fraction is meant to be a conservative average to reflect that some populations would be protected if indoors.

Table 2 Summary of Consequences

Release Type	Ignition Type	Consequence Type	Endpoint Metric	Fatality Fraction
Instantaneous	Immediate	Fireball	Fireball radius	0.7
Continuous		Jet Fire	Distance to 35kW/m^2	1
		Pool Fire	Distance to 35kW/m^2	1
Either Instantaneous or Continuous	Delayed	Flash Fire	Lower flammability limit (LFL)	0.7
		VCE	None	--
Escalating	Piloted	BLEVE	Distance to 0.3 bar	1

2.4.1 Fireballs

Fireballs are short-lived flames, generally resulting from ignition and combustion of turbulent vapor / two-phase fuels in air. They are usually caused by instant or short duration releases related to catastrophic failure of pressurized vessels or pipelines.

Persons indoors are generally protected from the short-duration radiation effects from fireballs. This framework applies a 0.7 fatality probability within the fireball radius to account for any people outdoors.

2.4.2 Jet Fire

The widely-used Cone model is applied as the basis for the jet fire modeling within Phast, which describes the shape of a jet flame as a frustum of a cone. The parameters describing the frustum, accounting for choked flow, are derived from comparisons with experimental data from laboratory and field tests. The key input parameters in defining jet fires are release rate, velocity, material, and release elevation.

This framework assumes all jets are oriented horizontally. A horizontal jet fire typically results in a larger hazard zone than a vertical or angled release and is generally more hazardous for personnel and equipment. The jet flame lengths and the subsequent radiation hazard ranges are primarily driven by the release rate and the material. Immediate fatality is assumed for all personnel within the 35 kW/m^2 radiation contour of a jet fire or a pool fire.

2.4.3 Pool Fire

The pool fire model in Phast calculates the shape and intensity of the flame, and a range of radiation results. A pool fire flame is modeled as a cylinder sheared in the direction of the wind, with diameter, height, and tilt angle (measured from the vertical). The flame shape gives input to the radiation calculations. The pool diameter is calculated by:

- Continuous liquid leak - The stable burning size is calculated, where the mass burning rate balances the mass release rate of liquid; the pool diameter is, then, the lesser of the stable burning size or the bund diameter if a bund is defined.
- Unbunded leak - If a bund area is not defined, the pool fire model takes into account any physical barriers to the spread of the liquid pool. As such, the pool is allowed to spread at a uniform depth until it attains a minimum thickness at a steady state. Factors such as sloping, drainage, and curbing in the immediate area are therefore not considered in determining the pool diameter.

The surface area of a pool is a critical parameter for fire calculations. Models are available for spills onto flat ground for both continuous spills (3) and instantaneous spills (4). Consequences from ignition of an "infinite" spreading pool are overly conservative and are not applied in the study.

2.4.4 Flash Fires

A flash fire is effectively the advancing flame front of an ignited vapor cloud. Although it presents significant personnel hazards (any outdoor personnel caught within the flash fire envelope are considered immediate fatalities), flash fires do not cause significant structural damage. There is little radiation outside of the lower flammable limit (LFL) contour, and damage done by the flash fire is typically restricted to ignition of easily ignitable materials such as cabling and plastic. Furthermore, flash fires do not generally create overpressures and as such their damage is limited to thermal impacts only. Persons indoors are generally protected from the short-duration radiation effects from flash fires. This framework applies a 0.7 fatality probability within the LFL contour to account for any people outdoors.

2.4.5 Vapor Cloud Explosions

The gas dispersion model within Phast requires as inputs: material, phase, release rate, duration, and velocity. Where the cloud is ignited without being in contact with any area of congestion, a *flash fire* is assumed to occur. The predicted overpressure caused by a VCE is associated with the volume (mass) of the flammable cloud confined within the obstructed region(s), which needs to be differentiated from the entire volume of the vapor cloud or the total released inventory.

While most roadways, railways, rivers and pipeline right of ways are not congested at levels comparable to industrial facilities, congested regions can locally exist along any transit way (tree or hedge rows, crowded railyards, tunnels). The dimensions, make-up and proximity of congested regions drive the predicted overpressure from a release. For simplicity, fatalities from VCE are excluded from this framework.

2.4.6 Escalation/BLEVE

BLEVE hazards are considered as escalation events. BLEVE scenarios require heat or flame impingement on a storage container to create the increase in internal pressure and weakening of the container structure. BLEVEs are considered for the rail and truck transport modes. If a railcar has an ignited hazard, it will likely impact a neighboring railcar which would eventually fail as a BLEVE scenario. If there is an ignited leak from the truck, then the scenario may escalate to a BLEVE scenario with the remaining inventory in the truck tanker.

2.5 Framework Limitations

The intent of the framework is to compare transport by mode. The framework does not include:

- Loading and offloading risk which are often risk drivers for barge, truck and rail transport modes. Loading and offloading hazards would be at the endpoints of the route and the risk impact would be localized to these areas and highly dependent on the local operations.
- Flow within a river channel and other effects of terrain. Hazard zones may be expanded by these elements; material carried downstream or channeled by ground terrain.
- Shrapnel. This is another potential hazard impact; but the potential zone is uncertain. Fatality estimates from the other hazards should be sufficient to form the basis of the framework calculation.

3 Transport Modes

3.1 Pipelines

The pipeline operation is assumed to be continuous and the loss of containment frequencies are on an annual basis. The framework risk estimate for the pipeline release is used as the base value and to normalize the other results; thus the pipeline release is independent of the volume transferred.

3.1.1 Frequency Source

Previous DNV GL evaluation (5) of U.S. Department of Transportation Pipeline and Hazardous Materials Safety Administration (PHMSA) pipeline incident records indicate that a 10-year average of liquid pipeline data (entire system) results in a leak rate of 2E-03 per mile per year. 90% of the incidents are a leak of the pipeline represented by a hole of 10% of the diameter. The remaining 10% is a rupture of the pipeline diameter. The pipeline size distribution indicates that the average pipeline carrying highly volatile liquids is roughly 14-inches in diameter. The frequencies presented in Table 3 are irrespective of population density.

Table 3 Pipeline Scenarios and Frequencies

Spill Size Category	Representative Release Scenario	Release Type	Frequency (per mile per year) by Population Density Category		
			Low	Medium	High
Leak	1.4-in release	Continuous	2E-3	2E-3	2E-3
Rupture	14-in pipeline rupture	Continuous	2E-4	2E-4	2E-4

The PHMSA estimated release frequencies are generic averages for liquid pipelines in the US. Although not applied in the current study, DNV GL has methods to assess pipeline route loss of containment frequencies highly refined by failure mechanism and local conditions (soil type, coverage, pipe thickness, etc.). This level of detail is not applied in the case study and framework as the goal is for concept comparison of transport modes.

3.1.2 Release Conditions and Consequence Sizes

Table 4 presents the estimated consequence hazard zones for the pipeline release scenarios. All hazard zones within the discussion are presented with respect to the downwind distance from the release location.

Table 4 Pipeline Consequence Zones

Spill Size Category	Downwind Distance of Consequence Zone (m)			
	Fireball radius	35kW/m^2	LFL	0.3 bar
Leak	--	80	90	--
Rupture	--	430	810	--

3.1.3 Ignition Probabilities

The TNO Purple Book is used to define both immediate and delayed ignition probabilities (6) for pipeline releases. Immediate ignition is often a result of the energy that caused the loss of containment and is a function of the reactivity of the material. TNO direct ignitions for liquids, 0.065 for both continuous and instantaneous releases, are applied for immediate ignition of pipeline releases.

Delayed ignition probabilities are a function of proximity and strength of ignition sources and their exposure to the flammable region of the cloud. For pipeline routes, the framework applies the TNO delayed ignition probabilities of 1.7E-4 per person per second to account for population. Total ignition probabilities are capped at 1.

3.2 Rail

Rail transport is assumed to be performed by unit trains, which means dedicated LPG rail tanker transport. The rail transport is assumed to consist of 120 railcars per transit. The rail transport volume within the 120 railcars is applied as the base volume to normalize the barge and truck tanker transport.

3.2.1 Frequency Source

A recent 10-year railroad accident history (2005-2015) published by the Federal Railroad Administration (FRA) Office of Safety Analysis is analyzed to estimate the railcar release frequencies (7). The FRA database updates the accident records monthly and specifies each railroad related accident by region, state, type of accident, type of track, track class, cause of accident, casualty subset, hazard material involved or not, and asset damage level. FRA data indicate that the frequency of having a railcar accident along the transit route is about 2.1 accidents per million transit mile.

Since additional precautions are taken when a rail car is carrying a hazardous material, like LPG, the fraction of accidents resulting in a release is estimated from a subset of the FRA data where hazardous materials are present. The probability of release given an accident is also a function of the number and type of cars involved in the accident. FRA data do account for the number of cars involved in the accident. When hazardous material is being carried, approximately 6.1% of transit accidents with 120 or fewer HAZMAT cars result in a hazardous material release. While accidents are less frequent during transit than in the yard, they are more likely to occur at higher speeds and, thus, more likely to result in a release given an accident. An overall release rate of 1E-07 per mile is used in the analysis.

The NRC has estimates of the breakdown of probabilities by spill size category and population density category (1) in Table 5. FRA data are also used to derive:

- The probability distribution for the range of potential release sizes for transit accidents and the size of the releases shown in Table 5 and Table 6, respectively.
- The number of cars derailed from the 3,159 transit accidents in the FRA database where there were at least 120 cars carrying hazardous material. Data show a maximum of 32 cars out of 120 cars releasing in such accidents (8).

Table 5 Rail Release Probability Distribution by Size and Population Density

NRC Spill Size Category	Probability by Spill Size (7)	Probability by Population Density Category (1)		
		Low	Medium	High
Minor	0.80	0.1	0.1	0.8
Moderate	0.12	0.3	0.4	0.3
Severe	0.07	0.6	0.3	0.2
Extra Severe and Extreme	0.01	0.9	0.05	0.05

Table 6 Rail Scenarios and Frequencies

Spill Size Category	Representative Release Scenario	Release Type	Frequency (per mile) by Population Density Category		
			Low	Medium	High
Minor	3-in release from fitting	Continuous	1E-08	1E-08	8E-08
Moderate	1 car rupture	Instantaneous	5E-09	6E-09	2E-08
Severe	6 car rupture	Instantaneous	5E-09	3E-09	2E-09
Extra Severe and Extreme	20 car and rupture	Instantaneous/ Escalating	1E-09	6E-11	6E-11

3.2.2 Release Conditions and Consequence Sizes

Table 7 presents the estimated consequence hazard zones for the rail release scenarios. The consequences are extracted for the ground level impact. Note the Minor spill scenario's LFL impact distance is small as it is modeled as an elevated release and only a small impact distance is estimated at grade.

Table 7 Rail Consequence Zones

Spill Size Category	Downwind Distance of Consequence Zone (m)			
	Fireball radius	$35 kW/m^2$	LFL	0.3 bar
Minor	--	50	1	--
Moderate	--	60	340	--
Severe	--	140	880	--
Extra Severe and Extreme	90	230	1630	65

3.2.3 Ignition Probabilities

The TNO Purple Book is used to define both immediate and delayed ignition probabilities (6) for barges, rail and trucks. Immediate ignition is often a result of the energy that caused the loss of containment and is a function of the reactivity of the material. LPG is classified as a K0 material as the flash point temperature is approximately -104°C. TNO road tanker direct ignitions (0.1 for continuous releases and 0.4 for instantaneous releases) are applied for immediate ignition of rail releases.

Delayed ignition probabilities are a function of proximity and strength of ignition sources and their exposure to the flammable region of the cloud. For rail routes, the framework applies the TNO delayed ignition probabilities of 6.7E-03 per second accounting for the proximity of the train and 1.7E-4 per person per second to account for population. Total ignition probabilities are capped at 1.

3.3 Barge

Barge transport would consist of a single riverboat pushing 2-3 tank barges.

3.3.1 Frequencies and Scenarios

A U.S. Department of Defense (DoD) study estimated in 1980 a barge accident rate of 9.66E-06 per mile with approximately 4% of accidents resulting in a loss of cargo (9). While this data is older, waterborne commerce rates along the rivers of the evaluated transit route have been relatively flat (10); so an overall release rate of 4E-07 per mile based on the 1980 accident data is used in the analysis.

The NRC has estimates of the breakdown of probabilities by spill size category and population density category (1) in Table 8.

Table 8 Barge Release Probability Distribution by Size and Population Density

NRC Spill Size Category	Probability by Spill Size	Probability by Population Density Category		
		Low	Medium	High
Minor	0.44	0	0.5	0.5
Moderate	0.48	0	0.9	0.1
Severe, Extra Severe and Extreme	0.08	0.1	0.9	0

Although the NRC suggests spill size probabilities, no indication is provided of the potential associated release size. The U.S. Coast Guard (USCG) keeps casualty data (11) along the river route which is applied in the case study. In the period between 2001 and 2011, there were 141 recorded casualties of tank barges, 27 (19%) of which resulted in discharged cargo. The amount of recorded cargo discharge ranged from 1 gallon to 18,700 gallons averaging 715 gallons. Based on DNV GL past experience with the review of historical incident databases, it is likely that there is an under-reporting of the no release events. So the frequency estimates from the DoD study are applied in the study instead of the USCG data.

Table 9 presents the definition of the representative scenarios and the associated frequencies.

Table 9 Barge Scenarios and Frequencies

Spill Size Category	Representative Inventory Released Scenario (gal)	Release Type	Frequency (per mile) by Population Density Category		
			Low	Medium	High
Minor	715	Continuous	0	9E-08	9E-08
Moderate	18,700	Continuous	0	2E-07	2E-08
Severe	672,000	Continuous	3E-09	3E-08	0

3.3.2 Release Conditions and Consequence Sizes

Table 10 presents the estimated consequence hazard zones for the barge release scenarios. The resulting hazard zones are idealized as there is no inclusion in the modeling made for river movement. Note that the moderate and severe spills have the same hazard zones; this results from the same discharge rate but different durations of the release event, and the reported hazard zones are the steady state impacts.

Table 10 Barge Consequence Zones

Spill Size Category	Downwind Distance of Consequence Zone (m)			
	Fireball radius	35kW/m^2	LFL	0.3 bar
Minor	--	20	15	--
Moderate	--	110	170	--
Severe	--	110	170	--

3.3.3 Ignition Probabilities

The TNO Purple Book is used to define both immediate and delayed ignition probabilities (6) for barges, rail and trucks. TNO road tanker direct ignitions (0.1 for continuous releases and 0.4 for instantaneous releases), are applied for immediate ignition of barge releases.

Delayed ignition probabilities are a function of proximity and strength of ignition sources and their exposure to the flammable region of the cloud. For barge routes, the framework applies the TNO delayed ignition probabilities of 5.0E-03 per second accounting for the proximity of the push boat and 1.7E-4 per person per second to account for population. Total ignition probabilities are capped at 1.

3.4 Truck

For truck tanker transport, there are required weight limits. Therefore, the inventory is dependent on the material; pure propane within a standard tanker truck would contain about 10,238 gallons at approximately 89% volumetric capacity. Roughly 395 trucks would equate to a 120-car train load.

3.4.1 Frequency Source

Data on vehicle-km was collected in the Vehicle Inventory and Use Survey, which is now discontinued. The latest exposure data is for 2007 (12). The survey data indicates an exposure of 800 million truck miles. The incident database (13) includes 14 releases of Class 2.1 gases from cargo tank motor vehicles in transit in 2007. This gives a leak frequency of 2E-08 per tanker mile.

The NRC has estimates of the breakdown of probabilities by spill size category and population density category (1) in Table 11. Table 12 presents the definition of the representative scenarios and the associated frequencies.

Table 11 Truck Release Probability Distribution by Size and Population Density

NRC Spill Size Category	Probability by Spill Size (7)	Probability by Population Density Category (1)		
		Low	Medium	High
Minor and Moderate	0.995	0.1	0.1	0.8
Severe and Extra Severe	0.004	0.3	0.4	0.3
Extreme	0.001	0.5	0.3	0.2

Table 12 Truck Scenarios and Frequencies

Spill Size Category	Representative Release Scenario	Release Type	Frequency (per mile) by Population Density Category		
			Low	Medium	High
Minor and Moderate	3-in release from fitting	Continuous	2E-09	2E-09	1E-08
Severe and Extra Severe	Tank rupture	Instantaneous	2E-11	3E-11	2E-11
Extreme	Tank rupture	Escalating	5E-12	3E-12	2E-12

3.4.2 Release Conditions and Consequence Sizes

Table 13 presents the estimated consequence hazard zones for the truck release scenarios.

Table 13 Truck Consequence Zones

Spill Size Category	Downwind Distance of Consequence Zone (m)			
	Fireball radius	35kW/m^2	LFL	0.3 bar
Minor and Moderate	--	110	360	--
Severe and Extra Severe	80	90	330	--
Extreme	80	--		55

3.4.3 Ignition Probabilities

The TNO Purple Book is used to define both immediate and delayed ignition probabilities (6) for barges, rail and trucks. TNO road tanker direct ignitions (0.1 for continuous releases and 0.4 for instantaneous releases), are applied for immediate ignition of truck releases.

Delayed ignition probabilities are a function of proximity and strength of ignition sources and their exposure to the flammable region of the cloud. For truck routes, the framework applies the TNO delayed ignition probabilities of 6.7E-03 per second accounting for the proximity of the truck and 1.7E-4 per person per second to account for population. Total ignition probabilities are capped at 1.

4 Transport Mode Comparison

The different transport routes are assessed for their respective fatal risk impact along their respective transit locations and distances in the case study. To generate the framework result, the per shipment volume risk is calculated.

Table 14 presents the framework result. The total maximum release frequency for the respective transport mode is presented. The pipeline frequency is an annual frequency per mile, while the rail, barge and truck values are per shipment per mile. The volume transported is indexed across the transport modes, where the pipeline total volume is assumed constant, and the 120 car rail train is used as the basis for the other modes. Thus, to achieve the same volume transported as the rail option, the barge has a multiplier of 2.4 and the truck transit has a multiplier of 395.

The societal risk impact for the respective transport modes is calculated and then normalized to the annual pipeline risk result on a per shipment basis. The rail result is 5 times greater than the pipeline. However, the barge and truck normalized per shipment risk are less than the risk from pipeline and rail. The barge risk is about 11 times lower than the pipeline. Note though that these results are by shipment and not total volume.

Figure 1 presents a graph of the risk results normalized to the Pipeline – Low Population Density risk result versus the annual volume shipped.

Table 14 Transit Mode Comparisons

Transport Mode	Release Frequency (per mile)	Multiplier for Volume Transported	Per Shipment Risk (normalized to annual pipeline risk)
Pipeline	2E-03 annually	1 (filled pipeline for a year)	1
Rail	1E-07	1 (120 car train)	5
Barge	4E-07	2.4 (2-3 barges per trip)	0.09
Truck	2E-08	395	0.4

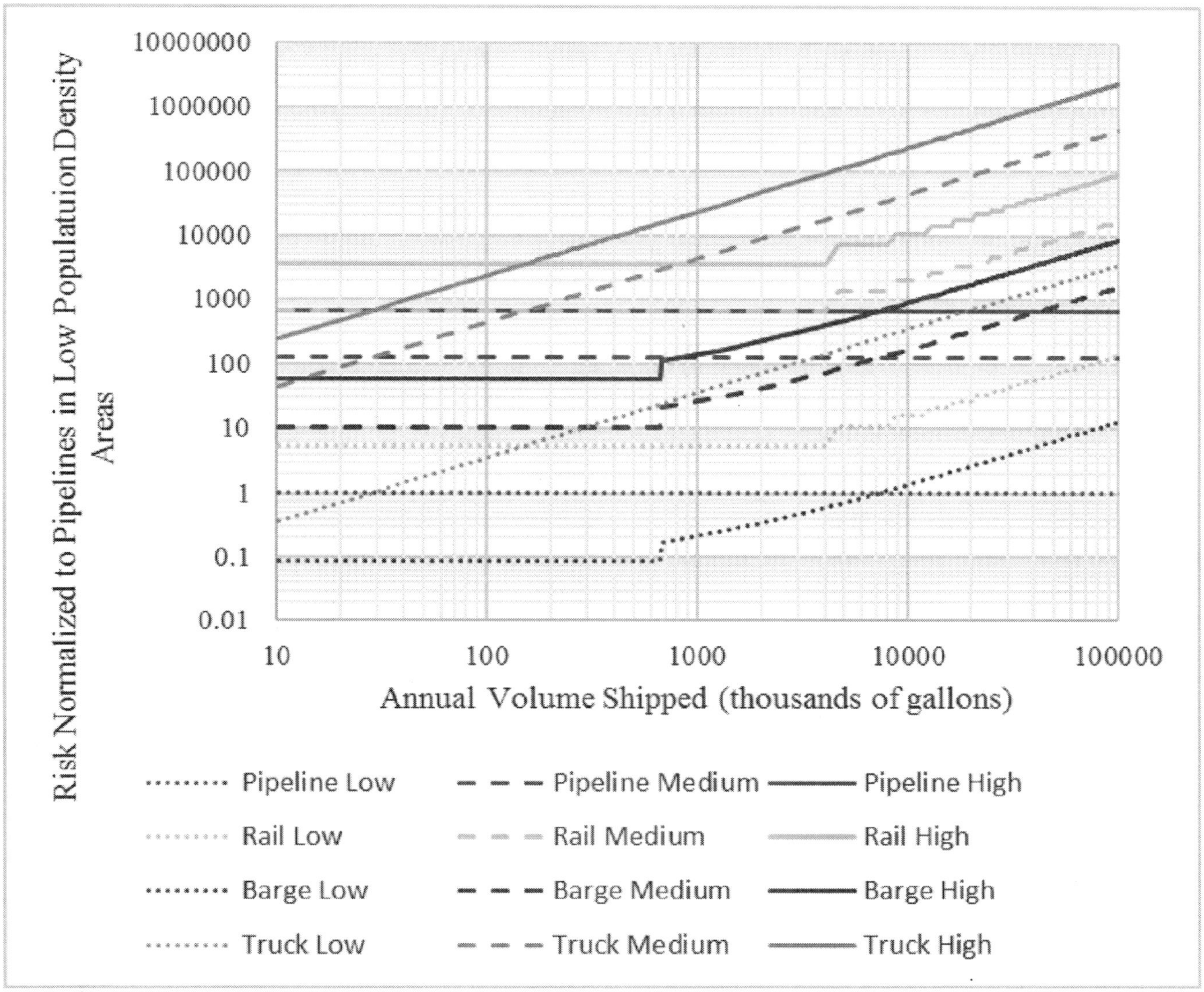

Figure 1 Transit Mode Risk Comparisons

Figure 1 presents the results for all four transit modes and three population densities. The following comparison discussions are made between the different transport modes assuming that the route length is the same across modes, and the population density is constant along the route length. However, both length and population density will vary by route even with the same end points. Route length will modify the risk linearly. Population density changes along the route has a more pronounced effect on the result.

4.1 Pipeline

As the results are normalized to Pipeline – Low Density, its curve is constant at value 1. As the pipeline can operate continuously, and the frequency is already on an annual basis, the pipeline results are constant across the x-axis of the annual volume shipped amounts.

As population density changes from Low to Medium to High, the pipeline result curves naturally increase in risk value. The Low to Medium result is a multiplier of 124; the Medium to High result is a multiplier of 5.

4.2 Rail

The rail result is constant up to the unit train 120 car volume, 4000 thousands of gallons. The slope of the risk result increases from the unit train volume as additional shipments are required to deliver the volume. The rail risk is always greater than the respective pipeline risk.

4.3 Barge

The barge result is constant up to the 700 thousands of gallons volume, and then the risk slope increases as additional transports are required to deliver the volume. The Barge – Low Density risk result is about 10 times less than the pipeline.

If the required volume to be shipped is in the order of 6700 thousands of gallons, which is 11 barge trips annually, then barge transport is the lowest risk option across the four modes and three population densities. However, if more than 11 barge equivalent volumes are required, pipelines present the lowest risk option.

4.4 Truck

The truck risk lines are constantly increasing with volume shipped given that additional truck shipments are required to achieve the volume. If only 3 or fewer truck shipments are required in a year, then the risk result is less than the pipeline option for the same volume equivalent.

5 Conclusions

The presented risk results vary across transport modes and population densities, and are presented on an annual volume shipped basis. Given an annual volume amount and a specified population density, comparison of the different transport modes is possible on a coarse basis. However, route length likely varies between the different transport modes and would factor the expected risk value. Population density would also likely vary along the length of a transport route and would change the expected risk result.

Generally, for shipment of volumes less than 7 million gallons annually, the risk impact to public from the different transport modes in increasing order is barge, pipeline, rail and truck. For shipment volumes over 7 million gallons annually, pipeline transport presents the least risk exposure of the four transport modes.

6 References

1. **U.S. Nuclear Regulatory Commission.** *Final Environmental Statement on the Transportation of Radioactive Material by Air and Other Mode.* 1977.

2. **U.S. Census Bureau.** [Online] 2010. [Cited: March 28, 2014.] http://www.census.gov/geo/www/gazetteer/files/Gaz_zcta_national.txt.

3. **Croce, Mudan K. S. and P. A.** *"Fire Hazard Calculations for Large Open Hydrocarbon Fires". SFPE handbook of Fire Protection Engineering,.* Boston, MA : Society of Fire Protection Engineers, 1988.

4. **Raj, P.P.K. and A.S. Kalelkar. Eds. D. H. Slater, E. R. Corran and R. M. Pitblado.** *Assessment Models in Support of Hazard Project. Major Industrial Hazards Project.* Sydney, Australia : Warren Center, University of Sydney, 1974.

5. *Using Past Incidents as Leading Indicators for Pipeline Integrity Management.* **Chen, Hao, et al.** Houston, Texas : American Institute of Chemical Engineers, 12th Global Congress on Process Safety, 2016.

6. **TNO Reference Manual .** *Bevi Risk Assessment Version 3.2, New "Purple Book". 2009.*

7. **U.S. Department of Transportation.** Federal Railroad Administration Office of Safety Analysis. [Online] 2015. http://safetydata.fra.dot.gov/OfficeofSafety/Default.aspx.

8. —. Federal Railroad Administration Office of Safety Analysis. [Online] http://safetydata.fra.dot.gov/OfficeofSafety/Default.aspx..

9. **Louis Berger & Associates, Inc. et. al.** *Final Environmental Impact Statement for the Safe Collection, Transportation, and Final Disposal of U.S. Department of Defense Stocks of DDT.* 1981. DACA45-80-0032.

10. **Department of the Army Corps of Engineers.** *Waterborne Commerce of the United States, Part 5, National Summaries.* 2016. CEIWR-WCUS-16-5.

12. **US Census (2009).** 2007 Economic Census: Transportation Commodity Flow Survey. [Online] http://www.rita.dot.gov/bts/sites/rita.dot.gov.bts/files/publications/commodity_flow_survey/2007/hazardous_materials/html/table_06b.html.

13. **Office of Hazardous Materials Safety (OHMS).** Incident Reports Database Search. [Online] 2014. https://hazmatonline.phmsa.dot.gov/IncidentReportsSearch/search.aspx.

AFPM/API Advancing Process Safety – Practical Tools for Process Safety Performance Improvement

Lara Swett
AFPM
1800 M St., NW Suite 900 North
Washington, DC 20036
lswett@afpm.org

Prepared for Presentation at
American Institute of Chemical Engineers
2018 Spring Meeting and 14th Global Congress on Process Safety
Orlando, Florida
April 22 – 25, 2018

AFPM/API Advancing Process Safety – Practical Tools for Process Safety Performance Improvement

Lara Swett
AFPM
1800 M St., NW Suite 900 North
Washington, DC 20036
lswett@afpm.org

Keywords: industry programs, process safety performance, process safety metrics, learning from incidents, lessons learned,

Abstract

American Fuel and Petrochemical Manufacturers (AFPM) and American Petroleum Institute (API) have been managing the Advancing Process Safety (APS) Programs since 2010. Through these programs, the associations have been able to collect a large amount of process safety data through API RP 754 collection, the event sharing database, and process safety site assessments. This paper will present the analysis of that data, the improvement opportunities identified, and the tools developed to assist sites in improving their process safety performance. One example of this is the identification of an industry opportunity to improve process safety events caused by open ended lines. Approximately 25% of process safety events have been caused by the failure to isolate energy during equipment commissioning, draining, and unloading/loading activities. AFPM in conjunction with several member companies have developed a suite of practice sharing tools called "Walk the Line" to assist our members in developing a program to address these types of events.

1. Background and Development of Advancing Process Safety Programs

The petroleum refining and petrochemical manufacturing industry is committed to protecting the health and safety of its workers, contractors and neighbors as well as the environment in which it operates. For decades, industry has worked together to develop safe work practices and standards aimed at reducing risks of major industry hazards and preventing process safety incidents. Despite these efforts, process safety incidents still occur.

In 2010, a group of refining and petrochemical industry executives belonging to the then National Petrochemical and Refiners Association (NPRA) now AFPM, tasked process safety professionals within their companies to develop a working group to explore ways to collectively improve process safety performance across the entire refining and petrochemical industry. The goal was to develop programs that would create a significant improvement in the collective industry process safety performance. Over a two-year period, the group developed a suite of programs. Over time, these programs have evolved and changed based on industry needs.

Today, the six programs are; Industry Learning and Outreach, Hazard Identification & Practices Sharing, Process Safety Regional Networks, Human Reliability, Mechanical Integrity, and Process Safety Site Assessment. Together these programs are called the Advancing Process Safety Programs (APS).

The objective of APS is to reduce process safety risk in industry by collaborating and leveraging industry knowledge.

This paper will discuss the process to collect and analyze process safety data being collected through APS, the observations from the data, and tools available to industry to address those observations.

2. Advancing Process Safety Programs

APS has a formal governance that sits between the two trade associations, AFPM and API. Each working group has a formal charter, operating guidance document, and accountability to the Process Safety Advisory group through the Process Safety Workgroup. This formal structure and oversight by a group of senior executives is one of the contributing factors to the long term sustainability and success of APS. Figure 1 shows the relationship between API and AFPM for APS, and the governance structure of the program.

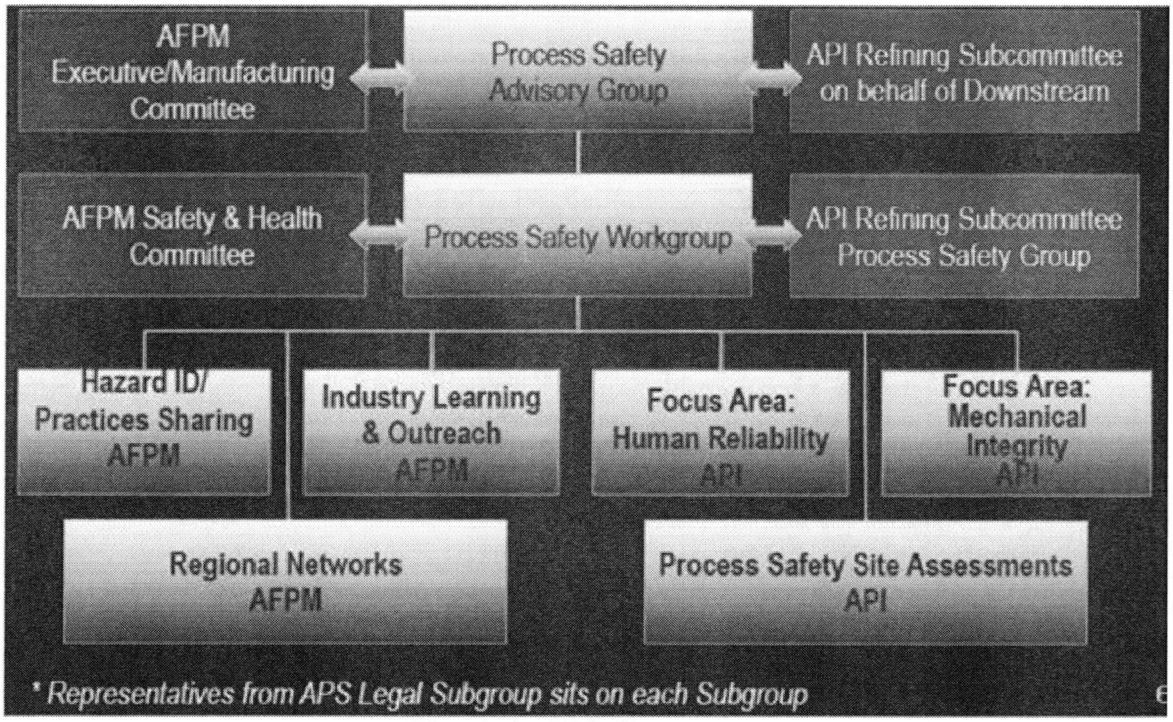

Figure 1: APS Governance

2.1 Safety Portal – Easily Accessible Tools

The Safety Portal[1] is an online database/website that houses all of the APS materials and is available to all AFPM members. It is searchable by keywords to allow for easy access to a variety of documents that address similar topics. The Safety Portal has been another key factor in the success of these programs. The site is currently undergoing a redesign which will make it more user friendly and will incorporate an improved search function.

The following document types are located in the Portal:

- Benchmarking application from *ANSI/API RP 754 Process Safety Indicators for the Refining and Petrochemical Industries*
- Event Sharing Database (Over 700 events)
- Hazard ID and Practice Sharing Documents
- Walk the Line Program page
- Industry Learning Reports
- Safety Bulletins
- Regional Network resources
- Presentations and Papers from AFPM Conferences

On average, there are about 2000 unique individuals that access the portal on a quarterly basis. This is one metric used to measure the program's success. Figure 2 shows the number of log-ins by unique visitors since its inception in 2012. Each individual is only counted once regardless of how many times they logged on. On quarter 4, 2017, we published several documents on a monthly basis and rolled out an outreach campaign that we believe led to a spike in log-ins.

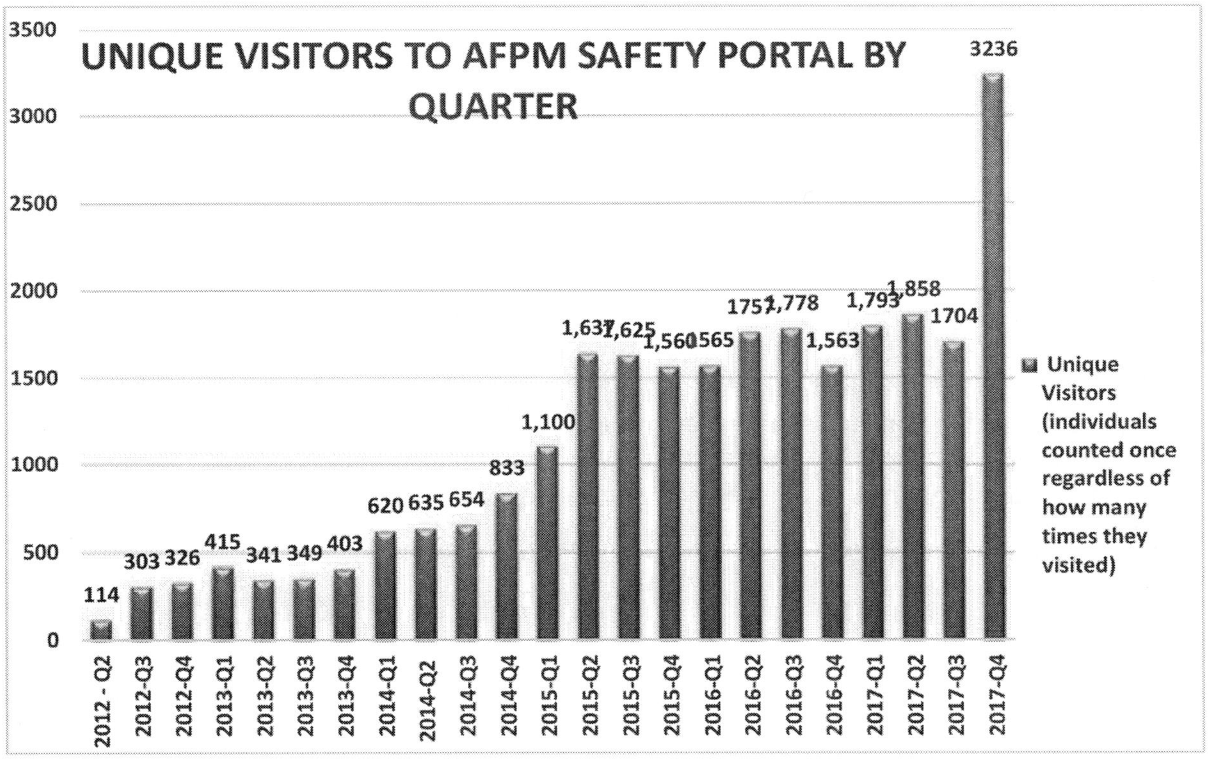

Figure 2: Unique Visitors to the Safety portal each Quarter

3. Advancing Process Safety Data Collection and Analysis

Through APS, AFPM and API member companies are asked to provide on an annual basis their Tier 1 and 2 Process Safety Events (PSEs) per *ANSI/API RP 754 Process Safety Indicators for the Refining and Petrochemical Industries* [2], to the event sharing database which includes cause and learning information and the option to conduct an API Process Safety Site Assessment [2]. Through the Industry Learnings and Outreach Subgroup, these data streams are analyzed and are the basis for the AFPM Learnings Report [3], outreach activities, and identification of opportunities for improvement.

3.1 Data Analysis

Included in calendar year 2016 data analysis are 688 event sharing submittals, approximately 3500 Tier 1 and 2 events (between 2012-2016 for refining and petrochemical), and results from 56 site assessments.

This analysis provides insight into opportunities for improvement that will be discussed later, however for discussion purposes each input, API RP 754, Event Sharing Database, and Site Assessment results are discreet data sets and therefore some observations might vary from data set to data set.

3.1.1 General Observations from 2016 Data Analysis

In general, key trends did not change from previous year's data analysis. However, with the addition of a more granular collection process that includes causal factors and a break out of types of normal operation (recent upgrade that provides additional trending opportunities) in the API RP 754 data, there is additional insight on point of release, cause and mode of operation observations.

AFPM has been collecting, API RP 754 Tier 1 and 2 data for the past 5 years. Figure 3 showing the 5-year process safety event incident rate trends for refining and petrochemical facilities. Over that time, there has been a downward trend in industry PSEs. Through additional analysis of refinery size, the small sites, < 190,000 bbls/day have achieved the most significant decrease in PSE rates. Antidotal information suggests parallels between the APS programs and their performance improvement. We do not have the same antidotal evidence on the petrochemical side.

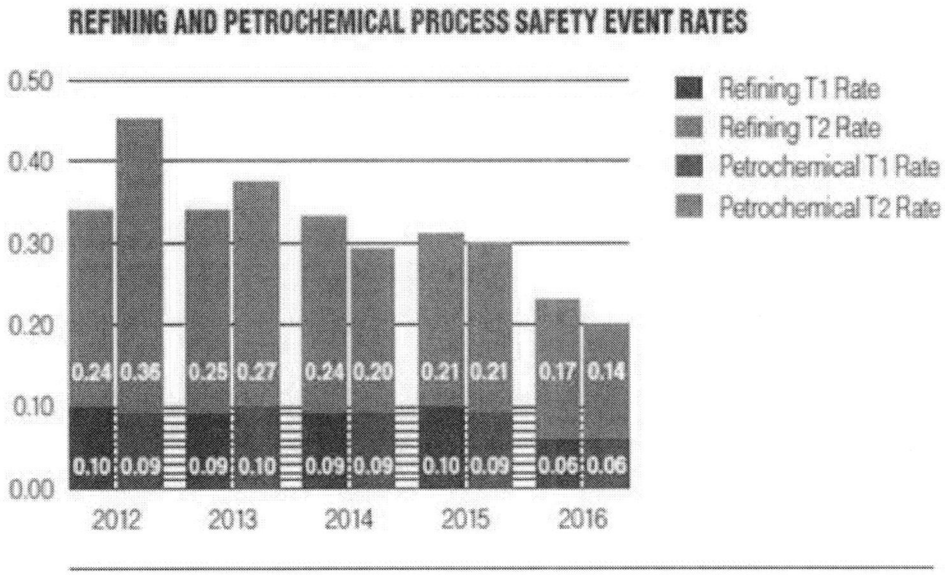

Figure 3: API RP 754 PSE trends for 2012-2016

Figure 4 illustrates where these Tier 1 and 2 releases are occurring in the Refining and Petrochemical industries. Combined, Piping Systems and Atmospheric Tanks categories comprise over half of the PSEs.

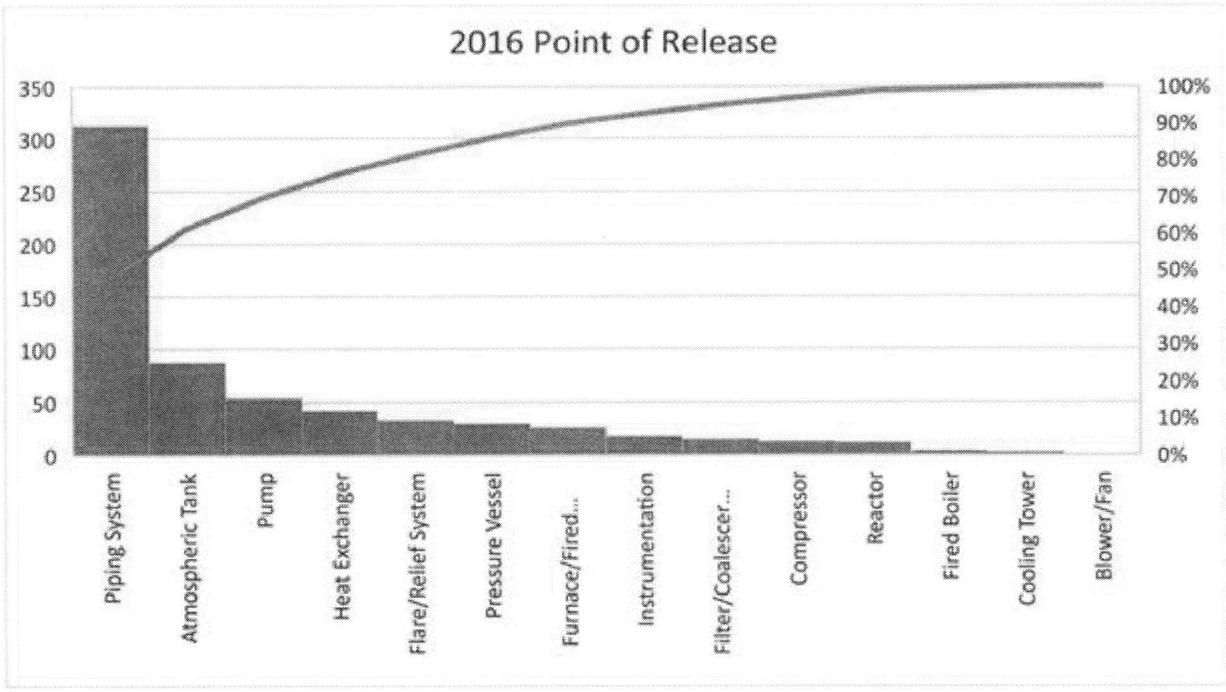

Figure 4: Equipment Point of Release, 2016 API RP 754 Data, Tier 1 and 2, Refining and Petrochemical

Upon further breakdown of that data:

Piping Systems - 43% of 2016 reported API RP 754 event data involved piping systems. Of those events, 57% were Large Bore Piping (>2 inch) and 42% were Small Bore Piping (<2 inch).

- Piping Leak Type:
  - 48% Piping Leaks
  - 24% Valve Leaks
  - 12% Flange/Gasket Leaks
  - 4% Tubing Leaks
  - 4% Fitting Leaks
  - 3% Open ended line
- Valve Leak Causes:
  - 54% Valve left open (Walk the Line Event)
  - 20% Valve Body Leak
  - 15% Valve did not hold/ leaked through

Atmospheric Tanks – 13% of events involved Atmospheric tanks, this percentage has remained consistent at 12-13% over the past 4 years. Forty four percent of PSEs related to atmospheric tanks were overfills and 87% of those events occurred during Normal Operating, Filling/Draining activities.

Of all the 2016 PSEs submitted, 59% occurred during Normal Operation and 50% of these events had at least one cause of Equipment Reliability and 62% had a cause of Human Error. Note, the submitter can choose more than one cause per event.

Human Error was the cause of 23% of all Tier 1 and 2 events submitted through API RP 754 data collection. Sixty two percent of those events occurred during Normal Operations. The following activities had at least one cause of Human Error:

- 83% Equipment Prep/Taking Equipment Out of Service
- 57% Loading/Unloading
- 56% Equipment Commissioning/Putting in Service After Maintenance
- 36% Filling/Draining
- 36% Steady State
- 25% Changing Line-ups

By looking at the Event Sharing database, we can observe some additional information that can provide additional information on causes:

- Fixed Equipment Mechanical Integrity – internal and external corrosion, erosion, cracking, and inspection less than adequate
- Equipment Reliability – premature failure and maintenance/repair less than adequate
- Human Factors – valves left open, open ended lines, loading, unloading, and tank filling

- Design – winterization, specs not adequate, followed by gaps in the specifications and construction

Further information can be pulled out from the Process Safety Site Assessment results. Figure 6 represents the results from the API Process Safety Site Assessments which show that the lowest scoring protocols are from Mechanical Integrity, specifically process hoses, integrity operating windows, and critical check valves. There is also a wide distribution of scores around conduct of operations topics like operator rounds and shift handover.

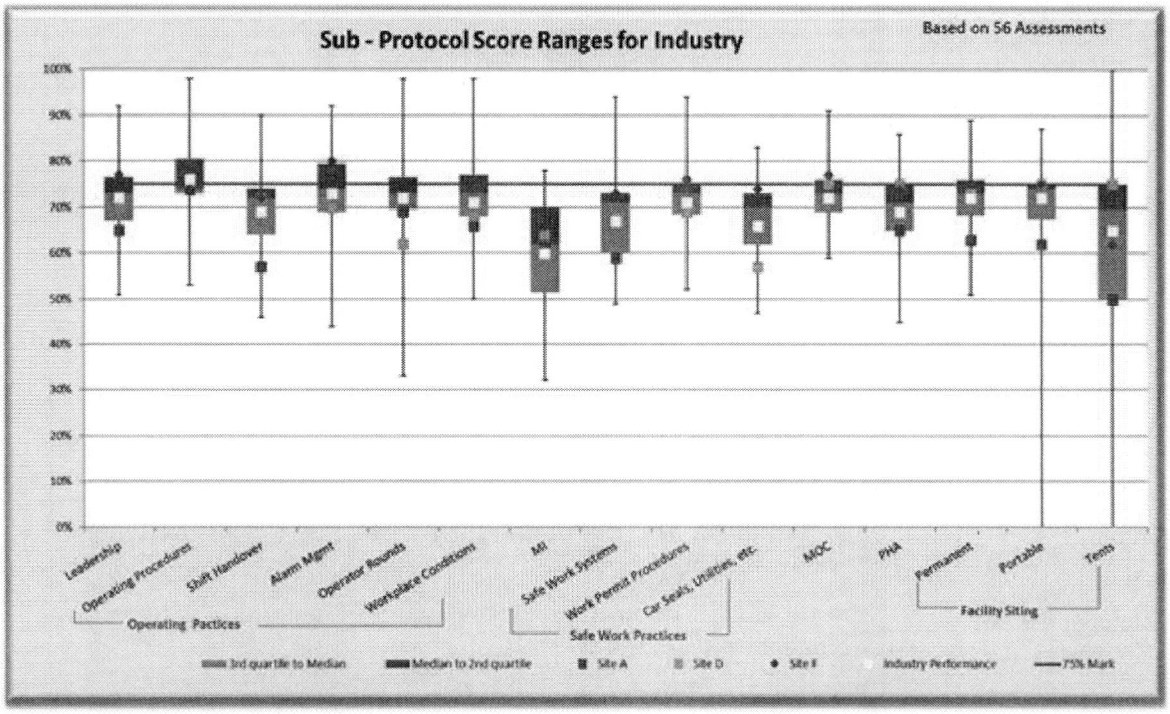

Figure 6: API Process Safety Site Assessment Score results based on 56 assessments

3.1.2 Data Conclusions

Over the past several years of data analysis, there have been two reoccurring themes among the reported causes; human reliability and mechanical integrity account for a majority of events that occur within industry. If you break down both categories a little further, human reliability as it relates to procedures is an improvement opportunity as well open-ended lines. On the mechanical integrity side, there are opportunities around operating integrity windows and critical check valves.

4. Opportunities for Industry Improvement

As part of APS, the trade associations have had the opportunity to work with the Legal Subgroup and develop several tools to enhance sharing and learning across the industry. These tools include;

- Safety Bulletins – A summary of process safety events with common causes
- Company Alerts – Companies share their alerts for broad distribution on a specific incident
- Practice Sharing documents – sharing the summary of a site's practice
- Hazard ID documents – a list of process safety hazards associated with a specific type of equipment or maintenance/operating activity
- Learnings Report – Annual summary of cumulative data analysis

Over the years of data review, several topics have been identified that have warranted additional sharing and outreach to industry. As mentioned above, the data points to a needed focus on topics related to Mechanical Integrity and Human Reliability. Below is a listing of tools developed to date, the section after will focus on the AFPM Walk the Line Program that is aimed to eliminate operator line-ups.

Table 1. APS Tools

Topic	Tools
Winterization and Dead legs	Hazard Identification: Winterization Hazard Identification: Dead legs
Management of Temporary Repairs	Safety Bulletin: Hazards of Installing Temporary Repairs for Piping Systems - Clamps Hazard Identification: Temporary Repair of Piping and Piping Components Practice Share: Temporary Repair Checklist Practice Share: Energy Isolation Procedure
Corrosion Under Insulation	Safety Bulletin: Hazards of Corrosion Under Insulation
Floating Roof Tanks	Safety Bulletin: Hazards of Floating Roof Tanks
Atmospheric Tanks	Hazard Identification: Atmospheric Tank Preparation for Out of Service Maintenance Hazard Identification: Atmospheric Storage Tank Operation
Piping Vibration	Safety Bulletin: Hazards of Piping Vibrations
Integrity Operating Windows	Practice Share: Integrity Operating Window Alarm Management

	White Paper on Integrity Operating Windows
Injection Point	Hazard Identification: Injection Point and Process Mixing Points
Hose Hazards	Hazard Identification: Hose Hazards
Critical Check Valve	Hazard Identification (in development)
Permit-to Work (Human Reliability)	Practice Share: Permitted vs Non-Permitted Work Practice Share: Standards for Work Permit Execution: Energy Isolation (LOTO); Joint Job Walks, Safeguard Verification & Stop Work Triggers Practice Share: Permit-Related Auditing Process Practice Share: Permit Issuer/Receiver Roles (Competence, Training & Execution)
Fired Heaters (Human Reliability)	Hazard Identification: Fired Heaters Practice Share: Fired Heater Risk Management Practice Share: Start-up Checklist Practice Share: Transient Operations HAZOPs Safety Bulletins: Cold Start-up, Hot Restart, Identifying and responding to bogged/flooding heater
Open-ended Lines (Human Reliability)	*See Walk the Line below*
Risk Assessment	Practice Share: Higher Risk Work Procedure Practice Share: HAZOP Vulnerabilities List
Other	Practice Share: Control and Defeat Review Practice Share: Radio Protocol

In addition, AFPM members have access to the Event Sharing database that contains over 700 process safety events to perform their own trending and analysis or to review when conducting a HAZOP or PHA.

4.1 Walk the Line an Industry Practice Sharing Program

Over the years, the concept of human reliability has been a reoccurring theme during the data analysis, many events pointed to some level of human factors that caused an event to occur. Specifically, a majority of the events were attributed to open valves during equipment commissioning, draining, and loading/unloading activities; many are "high risk" potential where human action is the last barrier to prevent. We have started referring to these events as Walk the

Line events and label them as such in the Event Sharing database. Based on the 2016 API RP 754 data analysis, 29% of events can be directly attributed to a Walk the Line causes.

Since 2013, a group working on this topic has developed an industry outreach and practice share program called, Walk the Line that is leveraged from a Celanese program that addresses the human reliability associated with conduct of operations and specifically, operator discipline. It reinforces the non-negotiable fundamental responsibility of an operator to know with 100% certainty where a material will flow between any two points in a process. Walk the Line addresses the culture of operator line-up and provides tools such as a hazard identification and practice sharing documents, workshops and checklists that help the operator become successful in meeting this responsibility. A paper describing Walk the Line was presented by Jerry Forest, Celanese, at the 2014 Process Plant Safety Symposium (PPSS). The paper and presentation won the annual PPSS Best Paper Award.

Walk the Line's philosophy on how to address human behavior

1. Change the Culture by raising awareness: Constantly set and reinforce the expectation to "Walk the Line" (know where a material will flow in a process). Tools provided include newsletters describing Walk the Line and a video that summarizes the concept.
2. Provide operational continuity and discipline tools to help the operators understand the current operating state of the plant. These include templates for shift instructions, shift notes, shift relief, operator reading sheets, and design practices.
3. Provide operational readiness tools to help operators be successful in line-up situations that have the highest probability of error including startup and returning equipment to service from maintenance. These tools include commissioning practices, Safe Operating Procedure design, and Pre Start-up Safety Review templates and practices.

Since the AFPM Walk the Line rollout in 2014, it is estimated that eleven companies have shared at least one practice and approximately thirty-five companies have implemented at least one aspect of Walk the Line.

AFPM has held a Walk the Line Workshop for the past two years with 200 people attending the 2017 Workshop. The 2018 Workshop is currently being schedule for, July 25-26 at the Waterway Marriott in the Woodlands, Texas. Contact the author for registration information.

All of these tools are available on the AFPM Safety Portal by searching Walk the Line or WTL.

Table 2. Walk the Line tools

Topic	Tools
Toolbox Topics:	Bleeders

	Coker Switch Instrument Lineup Barge Loading Wrong Valve Operated LOTO Error Maintenance Operations Handover Similar Valve Operation More Bleeders Line Open for Maintenance
Training:	Walk the Line Training Training for Front Line Supervisors Training for Secure Transfers Abbreviated Training
Hazard Identification:	Shift Turnover Maintenance Operations Turnover Operator Line ups
Practice Sharing:	First Line Break Point Labeling Open Valve Labeling and Management Refinery tank Farm and Terminal Product Transfer Shift Handover Evaluation Rounds Informal Unit Walk Through Operating Instructions Operator Shift Notes Shift Change Communications Shift Change Meeting Critical Bleeder Line Labeling Spring Loaded Valves Plug Program Checklist SOP Maintenance/Operations Turnover P&ID Walk down

	MOC Tagging Pre-Startup Safety Review Soap Testing IR Cameras Tracking Walk the Line Incidents
Presentations:	Walking Down the Line Barrier Analysis Technology to Support Standard Work Reliability-Centered Operations Rounds Operating Discipline and Walk the Line Reboot Flavor of the Month

One of the most notable successes of this program is how industry has adopted the terminology, Walk the Line, to identify similar types of events. Sites have posted signs, created hard hat decals, and even if they have not adopted any tools, they recognize the terminology and what it stands for. Based on anecdotal feedback, many sites that have adopted a form of Walk the Line have seen positive results.

5.0 Industry Outreach

Throughout the past eight years of working on APS, we have learned that continued outreach is a large factor in continued sustainability and industry engagement in the programs. For that reason, we have a designated Subgroup the responsibility for continuous outreach and industry education, the Learnings & Outreach Subgroup. To be added to the distribution list to receive information on upcoming webinars, documents, and events, please email Lara Swett at AFPM, lswett@afpm.org. You do not have to be an AFPM member to participate in webinars workshops.

5.1 Webinars

What started as a quarterly webinar for API 754 Question and Answer assistance has evolved into a webinar series that covers all elements of APS tools. For 2018, five webinars are planned. Each webinar will spend some time reviewing new published documents and tools, review API RP 754 questions, and also address one specific focus area such as Mechanical Integrity (April 10th), Human Reliability as it relates to Fired Heaters (June 19th) and Permit to Work (September 12th), and Regional Networks (November 13th). Contact the author for registration information.

5.2 Industry Presentations

The Learnings and Outreach Subgroup continues to look for opportunities to present the learnings from APS to a variety of audiences; senior leaders, process safety, maintenance and operations. The long-term goal of APS outreach is to share the learnings and available tools to disciplines broader than process safety as safety touches all disciplines.

5.3 Testimonials

One of the ways we measure effectiveness of the programs is by collecting testimonials from participants. The ultimate metric of successes is a reduction of process safety events across the industry which we are starting to see. However, from the testimonials below, it is clear that APS is having a positive impact on the industry at the site level.

"Site made changes on Energy Isolation procedures based on a discussion at a previous Regional Network Meeting."

"During shutdown and startup, this site pulls events from the safety portal event sharing database and shares with operators. They have also discovered learnings from some of these documents and as a result have made modifications to one of their units."

"Based on a presentation from a Regional Network meeting on high-risk work procedures, a site gathered new ideas to take back." (Note: document available in practices sharing section)

"Network participant shared the "Tank Preparation for Out of Service Maintenance" Hazard Identification document with a team of specialists at the site. The group integrated parts of the document with their site's procedures and work practices. Specifically, they leveraged information about roof support and tank cleaning from the document which resulted in an improved program."

"Took information shared at AFPM Regional Network on "Dangers of Corroded Tanks" back to management and will now inspect their tanks for microbial corrosion in the future."

"One company used the results from a site assessment, as well as the benchmarking results from the overall petroleum industry site assessments to drive a mechanical integrity subgroup at their company. This group presented data to the Board of Directors which resulted in the Board approving the hiring of a corporate MI Director to drive corporate MI initiatives."

"Company drafted a PHA checklist on Clamps/Temporary Repairs based on the Temporary Repairs Hazard Identification document."

"Took Safety Bulletin on "Overpressure of an Exposure-Proof Enclosure," and input that document into their company's High Learning Value Event process. The document was sent globally to all sites."

6.0 Conclusion

The AFPM and API have been working together on an industry led initiative, APS, to continuously improve process safety performance through enhanced information, communication and collaboration. What started as a commitment to share information broadly in order to improve all sites, has now become a movement within industry. Through the help of the Legal Subgroup, many barriers to sharing have been remove. This has helped companies feel comfortable sharing through APS, knowing they are helping another site or company. We are seeing improved process safety performance within the industry and there is antidotal evidence that APS has helped several sites and companies improve their performance. As the program matures, we will start reaching out to additional disciplines such as training, operations, and maintenance to further share learnings and tools. We will continue to measure effectiveness on an ongoing basis and are excited to see what this program can do for industry in another five years.

1 References

[1] AFPM Safety Portal, Available at https://safetyportal.afpm.org/ Accessed on March 20, 2018.

[2] American Petroleum Institute. (2016). *Process Safety Indicators for the Refining and Petrochemical Industries, (ANSI/API RP 754)*

[3] API Process Safety Site Assessment Site, Available at http://www.api.org/products-and-services/site-safety, Accessed March 20, 2018

[4] Forest, J. (2014, April) . Walk the Line. Paper presented at the meeting of the Center for Chemical Process Safety, New Orleans, LA

Optimization of extraction condition for phytic acid from peanut meal by response surface methodology

*Hong Ren *, Ting Li , Huijie Wan, Jinping Yue*

Beijing Technology & Business University, School of Food and Chemical Engineering, Beijing Higher Institution Engineering Research Center of Food Additives and Ingredients, Beijing Key Laboratory of Flavor Chemistry, Beijing, China

Abstract Phytic acid (PA), a molecule with high commercial value, is one of the important component in peanut meal. However, PA has not yet been isolated from peanut meal to exploit and utilize. This paper reported the extraction conditions of PA from peanut meal after removed protein. The independent variables were hydrochloric acid (HCl) concentration, solid to liquid ratio, extraction time and extraction temperature. Response surface methodology (RSM) was used to optimize the extraction conditions based on the extraction yield of PA. The results show that the second-order polynomial models derived from responses well with the experimental ($R^2 = 0.9783$). The optimal extraction condition was obtained with solid to liquid ratio of 1:16 (g:mL), HCl concentration of 0.02 mol/L, extraction time of 105 min, and extraction temperature of 30°C. At this condition, PA with higher purity were obtained. the extraction ratio was 6.12%, and the content of PA was 182.7 mg/g dry PA extract. The experimental values under optimal condition were in good consistent with the predicted values. The PA extracted from peanut meal was verified qualitatively by IR spectra. The extraction technology of PA from peanut meal has a strong potential for realized high-value utilization of peanut meal.

Keywords: Phytic acid, Peanut meal, Response surface methodology, independent variables, optimization

1. Introduction

Myo-inositol hexakisphosphate, also called phytic acid (PA), is considered as an anti-nutrient component because of its strong ability to combine with multi-charged metal ions, specially Zn^{2+}, Ca^{2+} and Fe^{3+} [1,2]. Recently, a variety of pharmacological activity of PA have been reported. S. Norazalina found the potential value of PA extracted from rice bran in reducing colon cancer risk in rats [3]. The finding showed treatment with 0.2% (w/v) PA extract gave the greatest reduction in the formation of aberrant crypt foci. Yukako Okazaki's study showed that PA may improve the composition of cecal organic acids, microflora, and mucins, and it may decrease the levels of serum proinflammatory cytokines in rats fed a high-fat, mineral-sufficient diet[4]. PA is a potential absorption enhancer of flavonoid components on tight junction integrity in Caco-2 cell monolayers[5]. PA is a naturally occurring constituent which exhibits protective action in Parkinson's disease and it has been shown to lower blood glucose levels[6,7]. Moreover, PA has been widely used in metallic material field as a new generation of green corrosion inhibitor for copper [8-11]. cupronickel B30 [12], brass [13] and Mg-Li alloys[14]. PA is widely used in food and light industry.

PA mainly distributes in seeds of plants, such as cereal grains, legumes, nuts, oilseeds, and so on, acting as a main source of phosphorus. Many cereal grains and oilseeds contain about 1~3% PA [15]. Peanuts is one of the most widely grown oil seeds in the world with 29 million tons produced every year [16]. In China peanuts are the most important oil seeds, ranking first among the world's peanuts producing countries[17]. Nowadays, a majority of peanuts are crushed to produce oil and this process generates a large amount of peanut meal as the by-product, which is about 9 million tons every year in china [18-20]. A large number of studies showed that peanut meal is a good source of active ingredients, which contains about 47~55% protein, 20~30% carbohydrates, 8~10% crude fiber, 2~3% fat, and 1.0~1.2% PA [21]. However, most of the available peanut meals are used as feed ingredient for animals, and only a small portion was used to recover proteins, causing great waste [22]. Therefore, obtaining the active ingredients and high-value utilization of peanut meal has become increasingly important task for effectively reducing peanut resources waste.

Compared with other plant materials, peanut meal is rich in PA, but there are no good process to isolate PA from peanut meal reported so far. The effective extraction procedure and method to separate PA from peanut meal are very necessary. In this study, we firstly removed proteins from peanut meal, and use response surface methodology (RSM) to optimize the extraction conditions. The objective of present study was to find the optimum conditions for the extraction of PA, including hydrochloric acid concentration, extraction temperature, extraction time and ratio of liquid to material, to maximize the extraction rate of PA from the peanut meal after removing proteins.

2. Materials and methods

2.1 Materials

Peanut meal was provided by Qinglonghu Wanchunyuan Vegetable Center in Beijing. The peanut meal was dried at 50°C in an oven after removing water-soluble proteins. The processing steps of removing proteins from peanut meal is shown in **Fig.1**. The dried peanut meal after removing proteins was ground using a mill.

Phytic acid sodium salt hydrate was purchased from Sigma-Aldrich Co. (St. Louis, MO, USA). Ferric trichloride was purchased from Tianjin Samtec Chemical Reagent Co., LTD. 5-Sulfosalicylic acid dehydrate was purchased from West Gansu Chemical Plant in Shantou city. HCl was purchased from China National Pharmaceutical Group Chemical Reagent Co., LTD.

2.2 Determination of PA content

The determination of PA content was evaluated by the Ferric trichloride combined 5-Sulfosalicylic acid method according to the procedure described previously[22]. 0.15g Ferric trichloride and 1.5 g 5-Sulfosalicylic acid was dissolved in 500 mL waters to be the Ferric trichloride and 5-Sulfosalicylic acid reaction solution. Inhale 0.0, 1.0, 2.0, 3.0, 4.0, 5.0 mL of 0.1 mol/L Phytic acid sodium salt hydrate solution in 6 colorimetric cylinders respectively, and trickle water into the colorimetric cylinders to 5 mL. A 5 mL Phytic acid sodium salt hydrate solution was combined with 4 mL Ferric trichloride and 5-Sulfosalicylic acid reaction solution. The absorbance of the solution was measured using a UV-vis spectrophotometer (754 UV-vis Spectrophotometer, Shanghai Jinghua Co, LTD) at a wavelength of 500 nm. Standard curve was draw with the absorbance as the ordinate and PA content as the ordinate. The regression equation was obtained as the following equation:

$$Y= -4.126x+0.878 \quad (R^2=0.9991) \qquad (1)$$

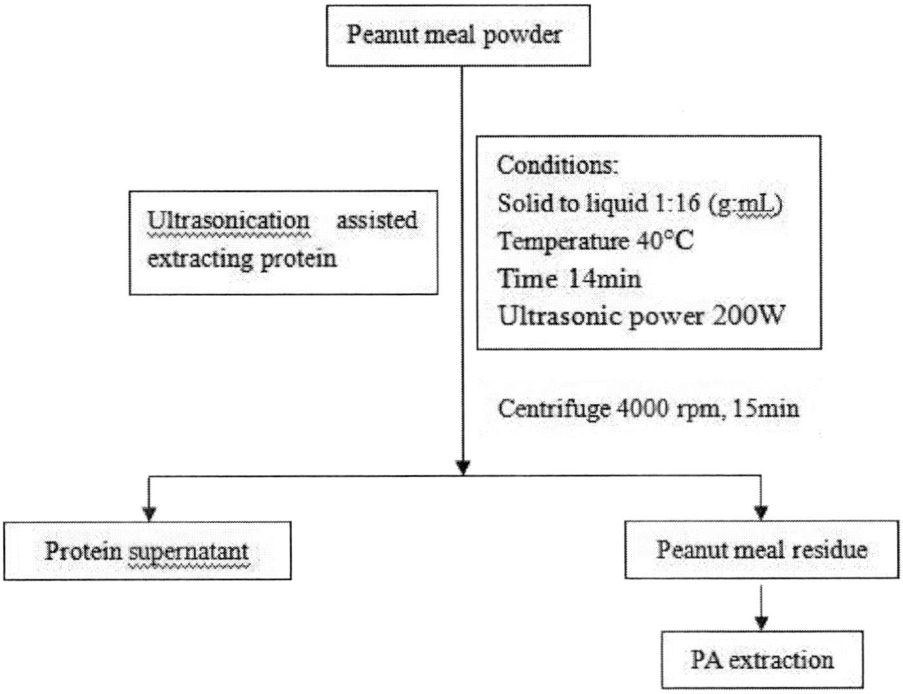

Fig. 1 The protein removal steps from peanut meal

2.3 Preparation of solid-liquid extracts

The influence of the solid-to-liquid ratio on the extraction of PA was investigated by using the following six ratios (1:10, 1:12, 1:14, 1:16, 1:18, 1:20; g:mL; sample powder: solvent). The hydrochloric acid concentration was fixed at the concentration of 0.01mol/L. The mixtures were extracted at 32°C for 83 min. The extraction solution was centrifuged at 4000 rpm for 10 min to obtain the supernatant. The content of PA (mg PA/g dry peanut meal) was then determined. The ratio that gave the highest value of yield was chosen for RSM.

2.4 Experimental designs

A d-optimal RSM experiment was designed using Design Expert (8.0.6, Stat-Ease Inc., USA). The independent variables were hydrochloric acid concentration (mol/L, X_1: 0-0.04), extraction time (min, X_2: 20-140), extraction temperature (°C, X_3: 25-60). The operating conditions were selected after single factor experiment and shown at **Table1**. The response variable was fitted by a second-order polynomial as follows:

$$Y = \beta_0 + \sum_{i=1}^{3} \beta_i X_i + \sum_{i=1}^{3} \beta_{ii} X_i^2 + \sum_{i=0}^{2} \sum_{j=i+2}^{3} \beta_{ij} X_i X_j \qquad (2)$$

Y : the predicted response; β_0: the intercept coefficient; β_i: the linear coefficient; β_{ii}: the squared coefficient; β_{ij}: the interaction coefficient; X_i, X_j: the coded independent variables; X_iX_j: the interaction terms; X_i^2: the quadratic terms.

The optimal conditions for the extraction of PA from peanut meal were then carried out using the equations of RSM.

Table 1 Independent variables and their levels used for Box-Behnken rotatable design

Independent variable	**Level**		
	-1	**0**	**1**
HCl concentration (X_1)	0.01	0.02	0.03
Time (X_2)	80	100	120
Temperature (X_3)	25	30	40

2.5 The extraction rate and purity of PA

The extraction rate of PA is the ratio of the weight of PA crude extract to the weight of peanut meal, and the purity of PA is the ratio of the pure PA to the PA crude extract. The calculation formulas of the extraction rate and purity are as follows:

The extraction rate (%) = (The weight of PA extract / The weight of peanut meal) × 100% (3)

The PA purity (%) = (The weight of PA / The weight of PA extract) × 100% (4)

2.6 Fourier Infrared spectrum analysis the PA of peanut meal

FT-IR measurement was recorded by a AVATAR 370 (Thermo Nicolet). 1mg dried PA sample was mixed with 200 mg KBr. The mixture was milled 5 min in the agate mortar and tablet. The spectra of each sample were used to analysis in the range of 400-4000 cm^{-1}.

3. Results and discussion

3.1 Determination of single factor experiment

3.1.1 Effect of solid/liquid ratio

In order to optimize the extraction of PA from peanut meal, solid to liquid had to be selected to study [23]. The ratio of solid to liquid varied from 1:5 to 1:20 (g:mL) while other extraction variables were fixed. In **Fig.2a**, as the decrease of the solid to liquid ratio, the content of PA increased. The highest yield of PA was found at 1:16 (g:mL). This indicated that more volume of solvent was required to sufficiently solubilize the target PA when the higher driving force of molecules was supposed to be higher in a high solid to liquid ratio. However, there was no significant differences in extraction yield from 1:16 to 1:20 (g:mL) ($p > 0.05$). Therefore, 1:16 (g:mL) was chosen as a compromise and fixed in the RSM experiment.

3.1.2 Effect of HCl concentration

In acid solutions, the complexing power of PA is declined, and the solubility of PA is increased. The effect of HCl concentration on the content of PA is shown in **Fig.2b**. The highest extraction yield of PA was observed at HCl concentration of 0.02 mol/L. From 0 to 0.02 mol/L, the content of PA increased. While from 0.02 to 0.04 mol/L, the content of PA decreased. As increased of HCl concentration, the PA was dissolved in the HCl solution because of the decline of complexing capacity of PA. When the HCl concentration was beyond the optimized concentration, other acid soluble components, such as acid soluble saccharide, amino acids, could be dissolved, may recombine with PA and form complexes, resulting in the decrease of the content of PA.

3.1.3 Effect of extraction time

The effect of extraction time on the content of PA was studied varying time from 20 to 140 min. The content of PA was substantially increased with the time rose from 20 to 100 min (**Fig.2c**). It had a decrease at longer time. It could be that other acid soluble components were dissolved and recombined with PA. The optimized extraction time was 100 min.

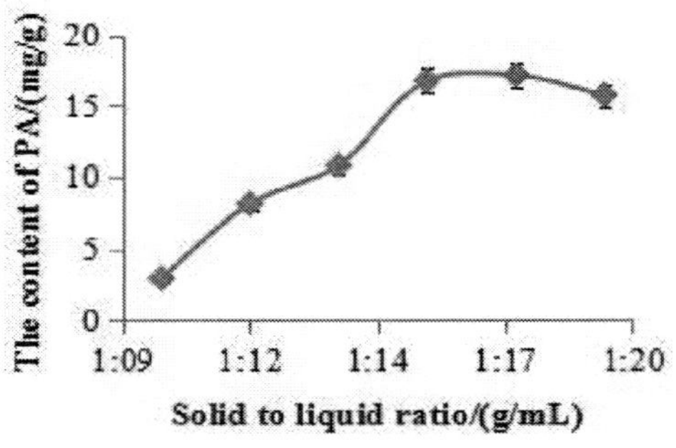

a

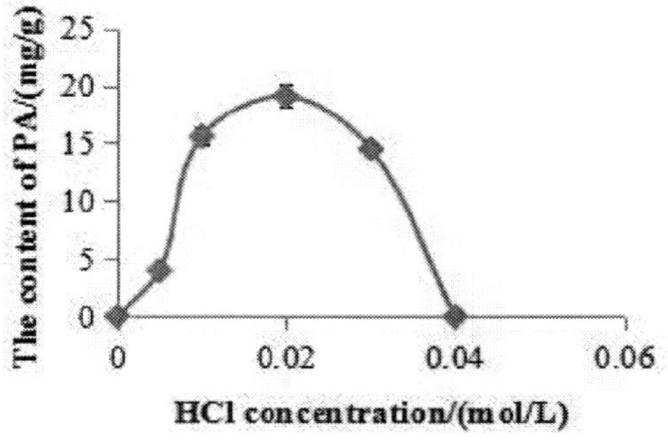

b

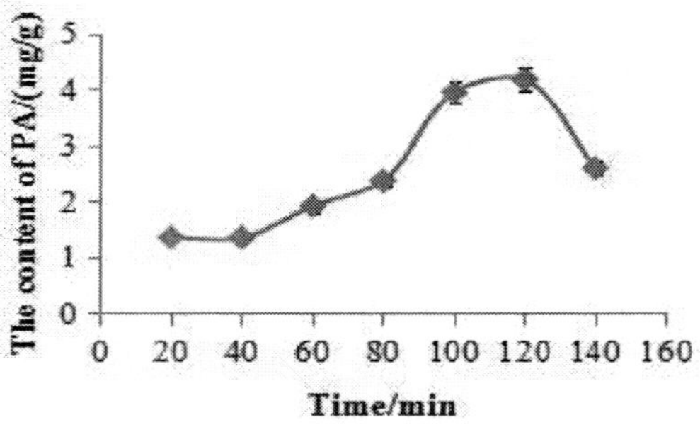

c

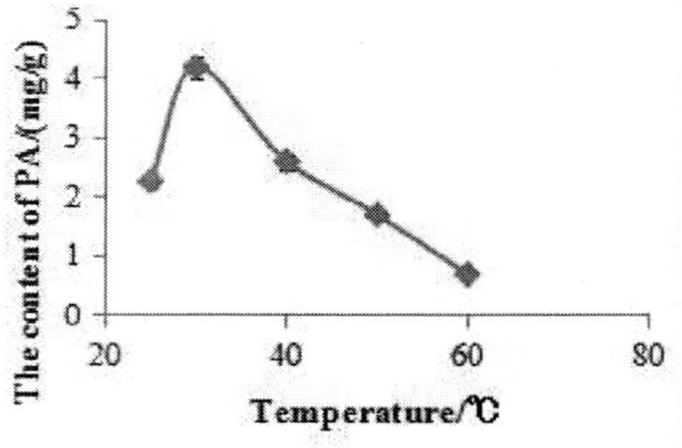

d

Fig.2 Single factor experiment

Effect of solid to liquid ratio (a), HCl concentration (b), extraction time (c) and Temperature (d) of PA from peanut meal. Different letters in the same sample indicate significant differences ($p < 0.05$) between treatments.

3.1.4 Effect of extraction temperature

As shown in **Fig.2d**, the yield of PA was increased with increased of extraction temperature from 25 °C to 30 °C. The maximum value of extraction temperature was at 30°C, having an obviously decrease with a further increase of extraction temperature. As the temperature increased, other ingredients were dissolved and combined with PA, causing the decreased of extraction yield.

According to the findings of above studies, the HCl concentration of 0.02 mol/L, extraction time of 100 min, extraction temperature of 30°C were selected as a central point for RSM. The ratio of solid to liquid was fixed at 1:16 (g:mL) in the following experiment.

Table 2 Coded levels, conditions runs and measured responses used in experimental design for response surface methodology

Run	X_1	X_2	X_3	**Yield (mg PA/g peanut meal)**
1	0.02	100.00	32.50	13.48
2	0.02	80.00	40.00	9.30
3	0.02	120.00	40.00	10.27
4	0.03	100.00	25.00	1.85
5	0.02	100.00	32.50	13.36
6	0.01	100.00	40.00	4.26
7	0.02	100.00	32.50	11.92
8	0.02	80.00	25.00	9.09
9	0.03	120.00	32.50	0.90
10	0.02	100.00	32.50	12.59
11	0.01	100.00	25.00	9.87
12	0.01	120.00	32.50	8.52
13	0.03	100.00	40.00	6.45
14	0.03	80.00	32.50	1.80
15	0.02	120.00	25.00	10.56
16	0.01	80.00	32.50	5.87
17	0.02	100.00	32.50	13.22

Table 3 Analysis of variance (ANOVA) for the fitted quadratic polynomial model for optimization of extraction parameters

Source	Sum of Square	Degree of freedom	Mean square	F value	p-value p>F	Significance [a]
Model	279.20	9	31.02	35.12	<0.0001	**
X_1	38.36	1	38.36	43.43	0.0003	**
X_2	2.20	1	2.20	2.49	0.1586	
X_3	0.15	1	0.15	0.17	0.6930	
X_1X_2	3.15	1	3.15	3.56	0.1010	
X_1X_3	26.05	1	26.05	29.50	0.0010	**
X_2X_3	0.062	1	0.062	0.071	0.7982	
X_1^2	173.67	1	173.67	196.60	<0.0001	**
X_2^2	20.78	1	20.78	23.52	0.0019	**
X_3^2	3.30	1	3.30	3.74	0.0945	
Residual	6.18	7	0.88			
Lack of Fit	4.47	3	1.49	3.47	0.1301	Not Significant
Pure Error	1.72	4	0.43			
Cor Total	285.38	16				

R^2=0.9783, Adj R^2=0.9505, Pred R^2=0.7401, adeq precision=16.526.

[a] *significant ($p < 0.05$), **extremely significant ($p < 0.01$)

3.2 Response surface analysis

3.2.1 Analysis of the adequacy of the fitted model

RSM was applied to determine the effect of the HCl concentration (X_1), extraction time (X_2), and temperature (X_3) on the content of PA (Y). The responses of each independent variable are listed in **Table 2**. The mathematical model for describing the content of PA was given by the following second-order polynomial equation:

$$Y=12.91-2.19X_1+0.52X_2-0.14X_3-0.89X_1X_2+2.55X_1X_3-0.12X_2X_3-6.42X_1^2-2.22X_2^2-0.89X_3^2$$

The response surface quadratic model was drawn up, and the analysis of variance for the fitted quadratic polynomial model for optimization of extraction parameters are presented in Table 3. The p-value for the model was less than 0.0001, which indicated the model was extremely significant and could be used to monitor the optimization. X_1, X_1X_3, and quadratic terms (X_1^2 and X_2^2) were extremely significant on the PA extraction value.

As **Table 3** shown that the "Lack of Fit" is not significant relative to the pure error. The coefficient of multiple determinations (R^2) for the quadratic regression model was 0.9783. The R^2 was higher than 0.7, indicated that the model was suitable for use in the experiment [24]. The value of adjusted determination coefficients (R^2_{adj}) was 0.9505, closed to the R^2. Therefore, this model can be used to navigate the design space [25].

3.2.2 Response surface and contour plots

The response surface plots and contour plots shown in Fig.3. It demonstrated the effect and interaction of independent variables on the yield of PA. Fig.3 a-1 and a-2 shows the effect of HCl concentration and extraction time on the content of PA. when the extraction time was fixed, the content of PA gradually increased with the increase of HCl concentration, up to the HCl concentration of 0.02 mol/L , a large increment of PA content was observed. This due to the increase of HCl concentration improved the solubility of the acid soluble components. When the HCl concentration was fixed, an obvious quadratic effect of extraction time was observed. Fig. 3 b-1, b-2, and Table 3 demonstrates the interaction of HCl concentration and extraction temperature is significant ($p < 0.05$). The content of PA as affect by HCl concentration and extraction temperature, exhibiting a clear increase in content with the raise of HCl concentration at a fixed extraction temperature. While an obvious quadratic effect of extraction temperature was observed. In **Fig.3** c-1 and c-2, a similar linear increase in content with the increase of extraction time ranged from 80 min to 104 min at a fixed temperature. While a decrease in temperature from 104 min to 120 min led to marked decrease in content of PA. This indicated that the extraction yield of PA depended on the medium level of time. These occurrences might be due to the recombination of PA and acid soluble proteins.

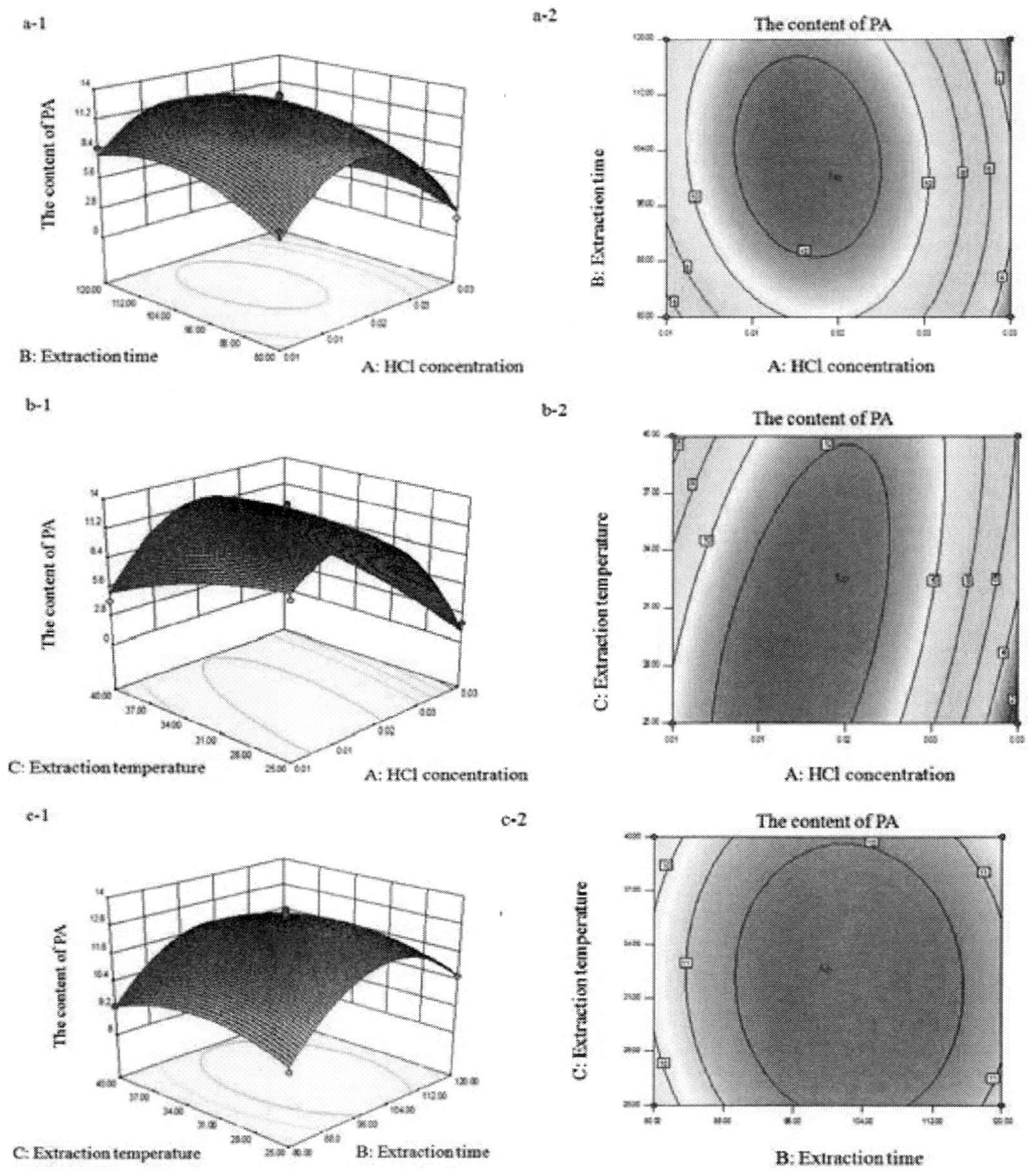

Fig. 3 Response surface plot and contour plot showing the effects of independent variables on the content of PA. a. HCl concentration and extraction time, b. HCl concentration and extraction temperature, c. extraction time and extraction temperature

3.2.3 Optimization of extraction craft parameters and the model validation

The optimal extraction conditions for PA were determined in order to simultaneously maximize the extraction. It was found that the optimal conditions were an HCl concentration of 0.02 mol/L, extraction time of 105 min, extraction temperature of 30 °C. The predicted and experimental values for the extraction of PA from peanut meal were 13.31 mg/g and 11.18 mg/g, respectively. In the optimal conditions, the experimental value was close to the predicted value, and the extraction ratio and content of PA was 6.12 ± 0.51% and 182.70 ± 2.35 mg/g, respectively. Removing the protein firstly and using the optimum extraction conditions can improve the extraction yield and purity. These results confirmed the predictability of the model to give the maximum yield of PA in the experimental process used.

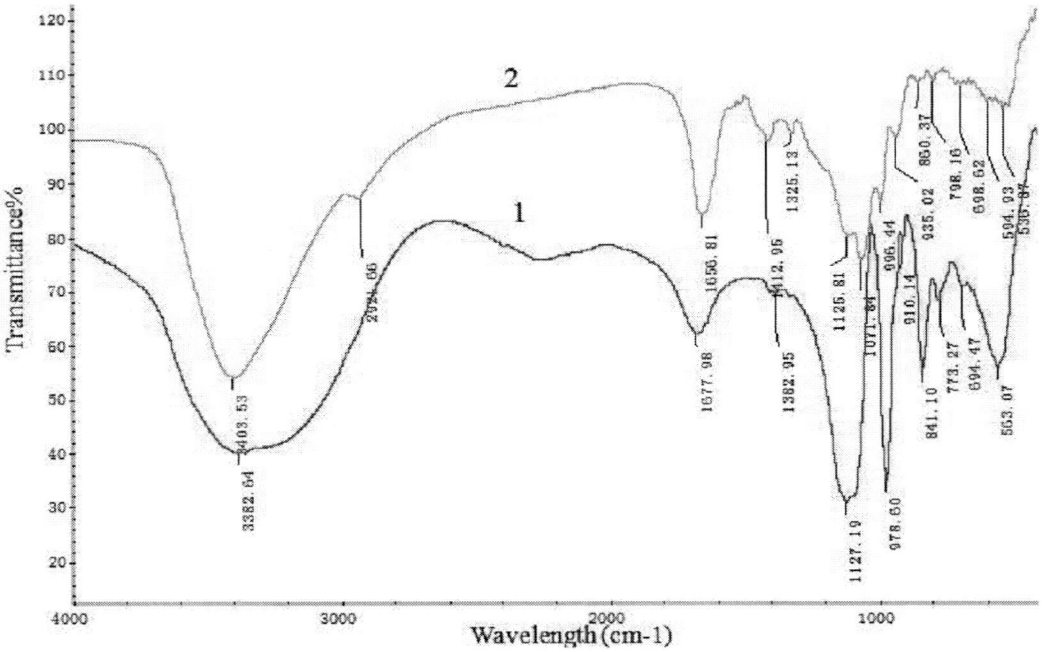

Fig.4 FT-IR spectra of PA extract.

1: Reference material phytic acid sodium salt hydrate from Sigma; 2: PA extract from peanut meal

3.3 Fourier Infrared spectrum analysis the PA of peanut meal

Each FT-IR spectra site of a compound provides qualitative information. PA contains phosphoester bond (P=O, P-O, C-O, the region of 1330~900 cm^{-1}; -OH, the region of 3650~3200 cm^{-1}). **Fig.4** was shown that reference material phytic acid sodium salt hydrate and PA extract from peanut meal have same peaks. Phytic acid sodium salt hydrate has the peaks at the wavelength of 3382 cm^{-1} (-OH), 1127 cm^{-1}, 978 cm^{-1} (P=O, P-O, C-O). PA extract has peaks at the wavelength of 3403 cm^{-1} (-OH), 2921 cm^{-1} (-CH), 1125 cm^{-1}, 1071 cm^{-1}, 996 cm^{-1} (P=O, P-O, C-O). Therefore, the PA extract from peanut meal was verified qualitatively by comparing with phytic acid sodium salt hydrate in the IR spectra.

4 Conclusion

In this study, we first removed proteins from peanut meal, and RSM was used to optimize the experimental variables including HCl concentration (mol/L), extraction time (min), and extraction temperature (°C). Solid to liquid ratio had significant effect on the extraction yield at the ratio of 1:16 (g:mL). This was chosen as the optimum parameter. The optimal conditions to obtain the highest extraction of PA were determined as follows: 0.02 mol/L of HCl concentration, 105 min of extraction time, 30°C of extraction temperature. At this condition, the extraction ratio was 6.12 ± 0.51%, which is far higher than the extraction ratio of traditional method, and the content of PA was 182.70 ± 2.35 mg/g dry PA extract. The PA extract derived from peanut meal could be verified qualitatively by IR spectra.

PA is the most important anti-nutrient ingredient in peanut meal. Peanut nutrient ingredients removing PA is one of the most attractive and promising sources of plant proteins or plant polysaccharides. On the other hand, peanut meal is the important renewable source of PA, PA extracted from peanut meal will be better applied in medicine and light industry. This study could not only be useful in science research and industry production of PA from peanut meal, but also conducive to high value utilization of peanut meal and reduce waste of peanut resources.

Acknowledgements

This work was supported by the National Key Research and Development Program of China (2016YFD0400802), National high technology research and development program of China (863 Program) (2011AA060701).

References

[1] Wang H., Zhou Y., Ma J., Zhou Y Y., Jiang H (2013). The effects of phytic acid on the Maillard reaction and the formation of acrylamide. Food Chemistry, 141 (1) , 18-22.

[2] Lopez H.W., Leenhardt F., Coudray C., Remesy C. (2002). Minerals and phytic acid interactions: is it a real problem for human nutrition? Int. J. Food Sci. Technol., 37, 727-739.

[3] Norazalina S., Norhaizan M.E., Hairuszah I., Norashareena M. S. (2010). Anticarcinogenic effcacy of phytic acid extracted from rice bran on Azoxymethane-induced colon carcinogenesis in rats. J. Experimental and Toxicologic Pathology, 62 (3), 259-268.

[4] Yukako O., Tetsuyuki K. (2014). Dietary phytic acid modulates characteristics of the colonic luminal environment and reduces serum levels of proinflammatory cytokines in rats fed a high-fat diet. Nutrition Research, 34 (12) , 1085-1091.

[5] Fu Q. X., Wang H. Z., Xia M. X. (2015). The effect of phytic acid on tight junctions in the human intestinal Caco-2 cell line and its mechanism. J. European Journal of Pharmaceutical Sciences, 80, 1-8.

[6] Lv Y. Q., Zhang Z., Hou L., Zhang L., Zhang J. Y., Wang Y. H., Liu C., XuP . P., Liu L., Gai X. Y., Lu T. X. (2015). Phytic acid attenuates inflammatory responses and the levels of NF-kB and p-ERK in MPTP-induced Parkinson's disease model of mice. J. Neuroscience Letters, 597, 132-136.

[7] Jin N. K., SungN. H., Kim H. K. (2014). Phytic acid and Myo-inositol support adipocyte differentiation and improve insulin sensitivity in 3T3-L1 cells. Nutrition Research, 34, 723-731.

[8] Notoya T., Otieno-Alego V., Schweinsberg D.P. (1995). The corrosion and polarization behaviour of copper in domestic water in the presence of Ca Mg and Na-salts ofphytic acid, Corros. Sci. 37 (1), 55-65.

[9] Yang H.F., Yang Y., Yang Y. H.(2005). Formation of inositol hexaphosphate monolayers at the copper surface from a Na-salt of phytic acid solution studied by in situ surface enhanced Raman scatteringspectroscopy Raman mapping and

polarization measurement. Anal. Chim.Acta, 548 (1-2), 159-165.

[10] Wang Y. .H., He J. B. (2012). Corrosion inhibition of copper by sodium phytate in NaOH solution: cyclic volt absorptometry for in situ monitoring of soluble corrosion products. Electrochim. Acta. 66, 45-51.

[11] Li C. C., Guo X.Y., Shen S. (2014). Adsorption and corrosion inhibition of phytic acid calcium on the copper surface in 3 wt% NaCl solution, Corros. Sci., 83, 147-154.

[12] Hao C., Yin R. H., Wan Z. Y., Xu Q J., Zhou G. D. (2008). Electrochemical and photoelectrochemical study of the self-assembled monolayer phytic acid on cupronickel B30, Corros. Sci., 50 (12), 3527-3533.

[13] Li Y., He J. B., Zhang M. (2013). Corrosion inhibition effect of sodium phytate on brass in NaOH media. Potential-resolved formation of soluble corrosion products. Corros. Sci., 74, 116-122.

[14] Gao L. L., Zhang C. H., Zhang M. L. (2006). Phytic acid conversion coating on Mg-Li alloy. Journal of Alloys and Compounds, 485 (1-2) (2009), 789-793.

[15] H. R. Park, H. J.Ahn, S. H.Kim, C. H.Lee, M. W. Byun, G. W. Lee. Determination of the phytic acid levels in infant foods using different analytical methods. Food Control, 17 (9), 727-732.

[16] Nader J., Fawaz N., Afif C., Louka N. (2016). A novel process for preparing low-fat peanuts: Optimization of the oil extraction yield with limited structural and organoleptic damage. Food Chemistry, 197, 1215-1225.

[17] Su G. W., Ren J. Y., Yang B. (2011). Comparison of hydrolysis characteristics on defatted peanut meal proteins between a protease extract from Aspergillus oryzae and commercial proteases. Food Chemistry, 126 (3) , 1306-1311.

[18] Wang Q., Shi A., Liu H., Liu L., Zhang Y., Li N., Gong K., Yu M., Zheng L. (2016). Chapter 5-Peanut By-Products Utilization Technology. Peanut: Processing Technology and Product Development, Copyright © 2016 China Science Publishing & Media Ltd. Published by Elsevier Inc., 211-325.

[19] Reddy N., Chen L.H., Yang Y. Q. (2013). Thermoplastic films from peanut proteins extracted from peanut meal. Ind. Crop. Prod. 43, 159–164.

[20] Wu H. W., Wang Q., Ma T. Z. (2009). Comparative studies on the functional properties of various protein concentrate preparations of peanut protein. Food Research International, 42 (3), 343-348.

[21] Zheng L., Ren J. Y., Su G. W., Yang B., Zhao M. M. (2013). Comparison of *in vitro* digestion characteristics and antioxidant activity of hot- and cold-pressed peanut meals. Food Chemistry, 141 (4), 4246-4252.

[22] Kiassos E., Mylonaki S., Makris D. P., Kefalas P. (2009). Implementation of response surface methodology to optimise extraction of onion (Allium cepa) solid waste phenolics. Innovative Food Science and Emerging Technologies, 10 (2), 246-252.

[23] Sahin S., Samli R. (2013). Optimization of olive leaf extract obtained by ultrasound-assisted extraction with response surface methodology. Ultrason Sonochem, 20 (1), 595-602.

[24] Wong K. H., Li G. Q., Li K. M., Valentina R. N., Chan K. (2017). Optimisation of Pueraria isoflavonoids by response surface methodology using ultrasonic-assisted extraction. Food Chemistry, 231 (15), 231-237.

[25] Bancha Y., Nuttapum S. (2015). Optimization of process parameters for phenolics extraction of Cratoxylum formosum ssp. formosum leaves by response surface methodology. Food Science Technology, 52 (1), 129-140.

512826 Influences of Geometric Factors on a Solar Thermochemical Reactor for Two-Step CO_2 Splitting

Abstract:

Solar thermochemical processes utilize the inexhaustible solar energy as thermal driving force to provide heat and motivate the reactions, which offer pathways to store solar energy to chemical fuels. Some elevated temperature reactions, like synthesis of syngas and methane reforming, are anticipating to accomplish in solar reactors, which have the ability to provide extremely high temperatures. Presently, the dominating factors in the design of solar reactors are heat and mass transfer with reaction mechanisms. A novel partition cavity-receiver reactor concept is proposed in this paper. For the purpose to provide a longer pathway of interaction between catalyst and reactants, a partition is introduced in this cavity-receiver reactor. A numerical computational fluid dynamics (CFD) analysis is performed to study on the influences of geometric factors, like length-diameter ratio, partition length and thickness, the transparency of partition, to fluid flow and heat transfer under constant solar power input for both two-dimensional and three-dimensional models. A two-step solar thermochemical redox reaction using ceria as catalyst to split CO_2 is modeled in the partition cavity-receiver reactor to investigate the relationship between geometric factors and reaction rates. Based on the comparisons and analysis of results, optimized geometric factors and corresponding operating conditions are discussed.

512849 Computational Tools for Designing Innovative Chemical Products

Abstract:

Designing new, innovative chemical products has always been a challenging task. Traditionally, the design process would require decades of product specific knowledge and a considerable amount of trial and error experimentation. Today we are developing a more systematic, scientific approach to chemical product design. New software tools can generate tens of thousands of candidate molecular structures and mixture formulations. Computational structure property relationships can be used to quickly predict the properties of each candidate and screen them against constraints on performance, safety and environmental impact. Promising candidates can then be screened using additional tools to ensure they can be produced in an efficient and sustainable manner.

We will discuss this new approach to chemical product design, demonstrate the use of computational tools for screening candidates and detail several examples. Specifically, we will:

1) Present an overview of the concepts behind computational chemical product design algorithms detailing how the computer can generate and enumerate â €œcorrectâ€ molecular structures.

2) Explain how group contribution estimation techniques are used to screen candidate chemicals.

3) Show how mixture estimation techniques can calculate the physical properties of candidate formulations.

4) Demonstration how basic chemical engineering knowledge is used to transform the often abstract goals of chemical product design into sets of well defined chemical and physical property constraints.

We will present several product design examples including the design of new refrigerants, solvent entrainers for extractive distillation, blends of lubricating oils, solvents for ink-jet printer ink and propellants for shaving cream.

513107 The Effect of Feed Compositon Variations from Upstream Plants on the Downstream Design Configruation of Natural Gas Liquid (NGL) Recovery Processes

Abstract:

Gas processing plant is subjected to variations in upstream conditions that has an impact on the performance of the in-operating plant. The feed composition determines what types of process configuration has to be implemented for the process. In this study, six representative NGL recovery processes were selected based on the criteria of their commercial availability, operating and capital cost savings. Each process scheme were built using HYSYS under common operating conditions for the purpose of comparison. The developed process schemes were tested for a range of various feed compositions classified as lean and rich. Economic evaluation were made by measuring the total annualized cost (TAC). The enhanced NGL recovery process (IPSI-1) showed the lowest TAC value compared to the others due to the presence of the self-refrigeration that reduces the external usage of propane refrigerant for the process. The trade-off that exist between the profitability and reliability of holding the process operating constraints were also been evaluated. The IPSI-1 scheme showed the highest profit margin with less than 6% risk of violation for the constraints.

513512 Modeling and Simulation of Diesel Hydrotreaters for ULSD Production: Selection of an Optimal Catalyst Composition

Abstract:

The required specifications for ultra-low sulfur diesel (ULSD) have driven refiners towards high performance hydrotreating catalyst that can efficiently remove difficult sulfur down to 15 ppm or lower in the final product. To select the best catalyst composition for the design of grassroots facilities or the revamp of existing ones, engineers need to account for the type of feed and the optimal operating conditions to reach hydrodesulfurization (HDS) and cetane number improvement targets.

This work focuses on the development of a flexible model for commercial ULSD hydrotreating units that predicts product properties, yields and hydrogen consumption for different catalyst compositions. The model uses a species set that contains the necessary structural information required to characterize different types of sulfur species as well as to distinguish among mono-/poly-aromatics, naphthenes and naphtheno-aromatics to model relevant reactions. The proposed reaction set accounts for desulfurization pathways by hydrogenolysis and hydrogenation, saturation of poly-aromatic, mono-aromatic and olefin species, cracking, and inhibition effects by nitrogen impurities in the feed.

Experimental data from the literature for CoMo and NiMo catalysts at different levels of desulfurization and aromatic saturation with temperature, pressure, and LHSV were used to determine two sets of kinetic parameters that represent each type of catalyst. Case studies were developed to assess the performance of each individual catalyst and combinations of catalysts in commercial ULSD hydrotreaters under different scenarios covering HDS and cetane improvement with limited hydrogen consumption, increasing amounts of light cycle oil in a straight-run gas oil feedstock blend, and low to high pressure operation from start of run to end of run conditions. For each scenario, an optimal catalyst composition was found, constrained to a final product sulfur specification of 15 ppm.

513598 Optimal Reaction Design of Alkoxy Silane Copolymers Using Design of Experiment Methods

Abstract:

Silicone-based copolymers such as polysiloxane and polysilsesquioxane have unique properties that make them well-suited for many industrial applications such as encapsulation and packaging of electronic components, anti-microbial coatings, and microelectronics[1,2]. They are particularly used in multi-layer lithography, a type of fabrication process for integrated circuits that provides increased resolution and sensitivity compared to other methods, and which relies on silicon-containing polymers to enable sharp pattern transfer. Integrated circuit applications require materials with very precise and tight quality specifications. To obtain these specifications, a good quantitative understanding of the molecular structures and properties of organosilicon polymers[3], which undergo structural changes during processing[3], is needed. Thus, understanding how defects in molecular domains influence the properties of the silicon-containing copolymers enables high-quality and controlled production scheme.

In this study, we applied the design of experiment methods (DOE) to gain a systematic understanding of the effect of different factors such as acid catalyst concentration, water to alkoxy silanes ratio, alkoxy silane monomers ratio, and condenser flow rate on the copolymer physical properties and structural chemistry. Here, poly(MTMS-co-TEOS), with the range of weight average molecular weight of 3000-6000 $gmole^{-1}$, was synthesized. The effect of reaction parameters and their interactions were studied through using different characterization methods such as TGA-IR, AFM-IR, NMR spectroscopies to achieve the optimal reaction condition for manufacturing products within the specification limits. It was found that increasing the ratio of condensation rate to hydrolysis rate and using a high ratio of trimethoxymethylsilane (MTMS) to tetraethoxysilane (TEOS) monomers induce a transition in the molecular structure and favors formation of random and cage structures, which can be attributed to defects, to linear ones. Moreover, we found out that acid addition rate has the strong influence on the formation of defects. We concluded that the formation of random and cage structures can be suppressed through increasing the condensation rate, improving the mixing rate, and controlling the addition rate of the acid catalyst.

References:

(1) Okoroanyanwu, Uzodinma. "Chemistry and lithography." Bellingham, WA: SPIE, 2010.

(2) Schmid, H., and B. Michel. "Siloxane polymers for high-resolution, high-accuracy soft lithography." *Macromolecules* 33.8 (2000): 3042-3049.

(3) Reichmanis, Elsa, et al. "Organosilicon polymers for microlithographic applications." *Advances in chemistry* 224 (1990): 265-281.

514041 Ethics As the Underpinning of Safety and Process Safety

Abstract:

Ms. Grubbe will discuss the various relationships between personal, professional and business ethics and the implementation of successful occupational and process safety efforts. She will pull data from known disasters, as well as examine the root causes of incidents and outline some "ethical failures." This examination will lead into the link between the corporate "tone at the top" and the underlying mindset of the executive team. As the leader of the AIChE Center for Ethical Practice and as a retired corporate executive, the speaker will be speaking from her four decades of experience.

514014 Multiple Ways to Failure - How Harassment Hurts the Engineering Profession

Abstract:

Deborah Grubbe will address the subject of harassment in the engineering workplace from the following points of view: 1) the corporate executive (male), 2) the male engineering or project manager, 3) the woman engineering executive/manager, 4) the woman engineer, 5) the woman engineer harassing the male engineer, and 6) the male engineer. She will address questions from her 40 years of experience as an engineer, a manufacturing and engineering executive, an HR manager, and a sponsor and participant of identity networks of all types. Deborah has served on the Corporate Partnership Council for the Society of Women Engineers and as a Board Member of the Women in Engineering ProActive Network. She has held a number of multinational executive roles and is currently a member of the National Academy of Construction. She is the current leader of the AIChE Center for Ethical Practice.

515420 Remarks on Ethics in Patenting

Abstract:

A patent applicant has a duty of candor in filing papers with the U.S. Patent and Trademark Office. This duty can implicate, for example, the need to be honest when presenting data, and particularly not falling into the trap of excluding less-favorable data when trying to demonstrate unexpectedly good results to the USPTO. There are some notable cases in which failure to be honest with the USPTO, such as by failing to present all the data, led to a patent being entirely invalidated as unenforceable.

513148 Refinery Front-End Crude Scheduling with Consideration of Unit Maintenance

Abstract:

The front-end crude scheduling (FECS) is very important for the entire supply chain management of refineries. It is used to tackle the crude movement problem covering crude vessel unloading at ports, temporarily storing at storage tanks, multi-oil transferring through long-distance pipelines, crude blending in charging tanks, and continuously feeding to refineries under specifications. Through FECS, an optimal crude movement schedule considering both profitability and operability can be identified. In previous studies, however, such a schedule was normally obtained under the scenario that all process units were under normal and deterministic operating conditions without the consideration of any possible abnormal situations such as the demand of certain unit maintenance. Note that the crude supply process is a highly unit-intensive and interactive process coupling both batch and continuous operations under heavy duties. Thus, the unit depreciation and performance degradation cannot be neglected, and maintenance should be regularly planned and conducted [1,2]. For instance, all equipment needs to be periodically checked, repaired and even replaced; otherwise, equipment failures may results in significant economic loss, and severe safety and environment problems. Some research papers have been published on various uncertainties during normal operations, such as shipping delays, demand change and tank malfunctions [3,4,5], these studies focused on the identification of post remedial measures for some uncertainty scenarios, instead of proactively considering unit maintenance operations in advance. Therefore, there still lack of FECS studies with the consideration of the planning of unit maintenance. In this paper, a novel FECS model combined with the planning of unit maintenance has been developed. It can provide optimal schedules for both front-end crude movement and maintenance operations (starting time and duration) toward the minimal operational cost. The efficacy and performance of the proposed model has been tested and demonstrated through various case studies.

References

[1] Amaran, S., Sahinidis, N., Sharda, B., Morrison, M., Bury, S., Miller, S., & Wassick, J., (2015) Long-term turnaround planning for integrated chemical sites. Computer and Chemical Engineering, 72, 145-158.

[2] Vieira, M., Varela, T., Povoa, A., (2017) Production and maintenance planning optimization in biopharmaceutical processed under performance decay using a continuous-time formulation: A multi-objective approach. Computer and Chemical Engineering, 107, 111-139.

[3] Xu, J., Qu, H., Wang, S., & Xu, Q., (2017) A new proactive scheduling methodology for front-end crude oil and refinery operations under uncertainty of shipping delay. Industrial & Engineering Chemistry Research, 56, 8041-8053.

[4] Li, J., Misener, R., & Floudas, C., (2012) Scheduling of crude oil operations uncer demand uncertainty: A robust optimization framework coupled with global optimization. AIChE Journal, 58, 2373-2396.

[5] Zhang, S., Wang, S., & Xu, Q., (2015) A new reactive scheduling approach for short-term crude oil operations under tank malfunction. Industrial & Engineering Chemistry Research, 54, 12438-12454.

513269 Process Stream Routing and Multi-Plant Operation of Industrial Wastewater Treatment Facilities

Abstract:

The authors have developed operating strategies for energy cost minimization and effluent regulation of an interconnected system of advanced wastewater treatment plants in Hillsborough County, FL. Inner-loop optimization has been performed on individual plant-wide transient operation of energy-intensive processes, while outer-loop optimization has been designed for influent and process stream manipulation and transmission between sites via plant-to-plant interconnects, transfer pumps, and lift stations. Forecasting models have been generated using artificial neural networks for inter-plant operability, and rules have been generated for supervisory decision making.

513168 Dynamic and Manufacturing Scheduling Approaching the Maximum Profitability Applications of Air Separation Units

Abstract:

Dynamic and Manufacturing Scheduling Approaching the Maximum Profitability Applications of Air Separation Units

Song Wang, Honglin Qu, and Qiang Xu

Dan F. Smith Department of Chemical Engineering

Lamar University, Beaumont, TX 77710, USA

Abstract

Cryogenic air separation processes use different in boiling points of the components to separate air into the desired products oxygen, nitrogen and argon. Unlike other chemical processes, the cost of feedstock (air) of an ASU is trivial; meanwhile, the energy supply of the entire process is totally in the charge of electricity power. ASUs are also characterized by fluctuating operating conditions to respond to changing product demands [1]. Although it is not much necessary to improve current ASU designs since there are quite proven techniques of cryogenic air separation design existing in industries; however, there are still significant potentials for improving their profitability in terms of optimal scheduling and operation ASU manufacturing.

Currently, many air separation plants are operated in a dynamic economic environment [2]. The electricity price could frequently fluctuate according to real time marketing (RTM) demands in different regions. In this paper, a general methodology based on dynamic simulation and scheduling of an ASU has been developed, which can optimize the ASU operation with respect to the electricity cost to achieve the best profit as well as highly efficiency. It contains three merits: (i) considering the entire ASUs dynamic simulation case study with detailed control strategy; (ii) optimal scheduling for maximize process profitability under uncertainties; (iii) smart production decision based on marketing supply-demands relationship. First, the methodology develops reliable dynamic simulation models under various production modes (capacities) and studies the dynamic transition performance among those production modes. Second, the optimal scheduling and transition will be conducted to maximize the plant profitability under the electricity cost pattern. Finally, the optimized ASU schedule and operations will be fine turned and validated by rigorous dynamic simulations. The developed methodology can be further extended to help ASUs for the predictive and proactive smart manufacturing.

Keywords: Air Separation Unit, Production Scheduling, Maximum Profitability, Smart Manufacturing

References

[1] Rui, H; Victor M. Zavala; Lorenz T. Biegler *. Advanced Step Nonlinear Model Predictive Control for Air Separation Units. Journal of Process Control. www.elsevier.com/locate/jprocont.

[2] Li, T*; Thierry, R; and Marc, B. Production Scheduling of Air Separation Processes. Air Liquide, Newark, DE 19702; Air Liquide, Champigny-sur-Marne, France; Air Liquide, Madrid, Spain.

513182 Retrofit Strategy for Heat Exchanger Networks

Abstract:

Numerous heat exchanger network (HEN) retrofit design methods have been proposed over the years, either based on topology modifications, heat transfer enhancement or a combination of both methods. However, selecting the most cost-effective retrofit strategy for a given network remains a great challenge.

In this work, the right balance between complexity of the retrofit solution, cost and benefit regarding energy recovery is sought. As such, the proposed retrofit strategy considers the application of heat transfer enhancement first, which is a low-cost retrofit option. Then structural modifications such as resequencing, stream splitting, adding a new match to create a loop or a path are explored in a step-wise approach. Finally, the third option (once enhancement and structural modifications are explored), combines the benefits of enhancement and modifications to the network to attain high energy recovery at reduced capital investment.

With heat transfer enhancement, a systematic retrofit methodology will be presented. The methodology can evaluate the feasibility of applying heat transfer enhancement techniques for a given HEN, identify the best location to apply enhancement, deal with downstream effects from the application of enhancement, and consider the effects of enhancement on pressure drop.

In general, a limitation with the application of enhancement is its effect on pressure drop together with the constraint in the degree of enhancement a heat exchanger can provide. These two factors influence the energy recovery that can be achieved. Therefore, in cases where high energy recovery is required, the next best retrofit strategy is to explore performing structural modifications. This work presents a new step-by-step retrofit approach (Pinch Retrofit Method) for performing structural modifications. This new approach is based on analysing the key features of an existing network as such providing insights into identifying the best series of structural modifications for a given network. This ensures that fewer modifications are required to achieve a degree of energy saving, and as a result is more cost-effective.

The next retrofit strategy considered is one that combines the benefits of heat transfer enhancement and modifications to the network. This can be achieved either by performing structural modifications and using heat transfer enhancement as a way of reducing the cost required (in terms of additional area required) or, replacing existing exchangers with plate exchangers (e.g. Compabloc). The use of plate heat exchangers is explored in this work as it has the benefits of improved energy efficiency, reduced fouling tendency and minimal plot space compared to shell and tube exchangers.

The methods discussed for retrofit are applied to an industrial case study. Pre-assessment on the case study identified a potential of reducing the furnace energy consumption by ~18%. The network minimum approach temperature is ~34°C, but the network is constrained with an exchanger minimum approach temperature of ~3°C. Applying the retrofit strategy proposed in this work, the best retrofit option was identified, which provided the right balance between complexity, cost and benefit. The results from analysis showed that significant energy recovery can be obtained with a payback of ~2 years.

Identifying the Causes of Acute Fouling in Refinery Preheat Train

Emilio Diaz-Bejarano[1], Francesco Coletti[1,2*], and Sandro Macchietto[1,3]

[1]*Hexxcell Ltd., Innovation Hub, Imperial College - White City Campus, 80 Wood Lane, London W12 0BZ, UK.*

[2]*College of Engineering, Design and Physical Sciences, Brunel University London, Uxbridge, UB8 3PH, UK*

[3]*Department of Chemical Engineering, Imperial College London, London SW7 2AZ, UK*

* f.coletti@hexxcell.com

Abstract

Crude oil fouling is typically treated in the literature as a phenomenon involving the deposition of organic material on the thermal surfaces of heat exchangers. Little attention is paid to the effects that inorganic species have on the thermal and hydraulic performance of the heat exchangers affected. When deposition of inorganic species occurs, it is often the cause of an upset in operations (e.g. issues with the desalter) that results in episodes of rapid (acute) fouling. In this paper, a method used to detect acute fouling episodes and identifying their causes is presented.

Introduction

Fouling deposition on heat transfer surfaces is a costly, long-standing problem particularly severe in pre-heat trains of crude distillation units [1]. Experimental analysis of fouling deposits collected upon dismantling such heat exchangers shows that both organic and inorganic species are typically present. These vary in proportion depending on the crude slates processed, refinery operating conditions, heat exchanger metallurgy and dominant fouling mechanisms. It has also been shown that the heat transfer given by fouling depends on the amount of the inorganic and organic phases, their structural arrangement and stratification or layering.

This paper presents a novel advanced mathematical model, implemented in Hexxcell Studio™, capable of describing the fouling deposit layer as a multi-component structure [2]–[4]. It is then shown how this approach is promising to detect acute fouling episodes and provide guidance on possible causes related to the composition of the deposit.

Method

Plant measurements are analysed using Dynamic Data Analysis™. This methodology uses an advanced solution mode of the shell-and-tube heat exchanger model implemented in Hexxcell Studio™ [5] to calculate the properties of the deposits over time as a function of inlet conditions of temperature and flowrate, heat duty and pressure drop. This allows estimating the tube-side deposit thickness and *apparent* thermal-conductivity over time (if shell-side fouling is negligible). The apparent thermal-conductivity is defined as the average value of the thermal conductivity of a homogeneous deposit that explains the overall thermal behaviour of the exchanger. The apparent thermal-conductivity is equal to the local conductivity of the deposit only if it remains constant over time, as this implies that the material that is depositing is of the same type and its properties do not change once it is settled. In reality, however, deposits are heterogeneous structures, with sub-layers that reflect variations in the deposition history (changes in type of foulant, temperature history, etc.). A heterogeneous deposit can be detected if the apparent thermal-conductivity varies over time. The pattern observed in the time-varying signal can be used to establish the underlying causes.

Case Study

The case study focuses on a double-shell heat exchanger located downstream of the desalter in a pre-heat train of a middle-eastern refinery. The full details of the case study are provided in [6]. Temperatures, flowrates and tube pressure drop data were available for a period of over three years in-between shutdown. The compositional analysis of the deposits, collected at the end of the run, were also available.

Results

The plant data analysis allowed calculating the time-profiles of deposit thickness and apparent conductivity, shown in Figure 1 and Figure 2, respectively. The deposit thickness profile showed a falling rate shape, with an evident acute deposition episode at around 150 days. The thickness at the end of the period was of ca. 2.75 mm.

The apparent conductivity showed variations over time, indicating the presence of a heterogeneous deposit layer. The trend shows an increase during the initial 175 days, followed by a gradual, although non-monotonic, decrease until the end of the operation period. In order to explain such behaviour, the deposition processed was postulated to follow this sequence of events: i) deposition of fouling of deposits with intermediate conductivity until the acute episode; ii) fast deposition of high conductive material during the acute episode; iii) after then, slow deposition of material with low thermal conductivity, with some occasional deposition of high conductive material.

In this context, taking into account the location of the exchanger and operating conditions (just downstream of desalter, temperature < 180°C), low and high conductive deposit can be associated to deposition dominated by organic and inorganic materials, respectively. An intermediate conductivity could be due to the two types of foulant depositing in similar proportions.

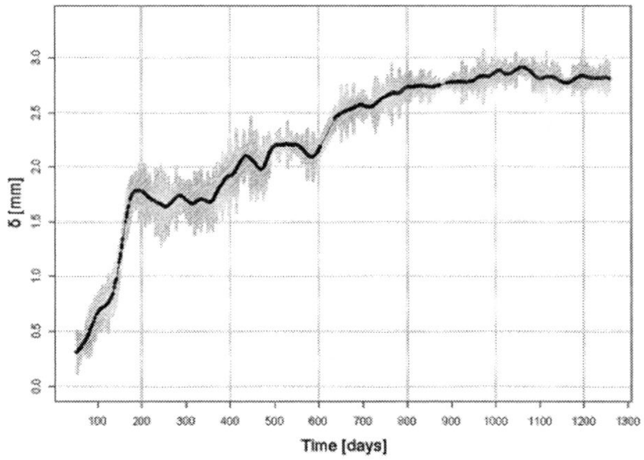

Figure 1. Deposit thickness over time (adapted from [6])

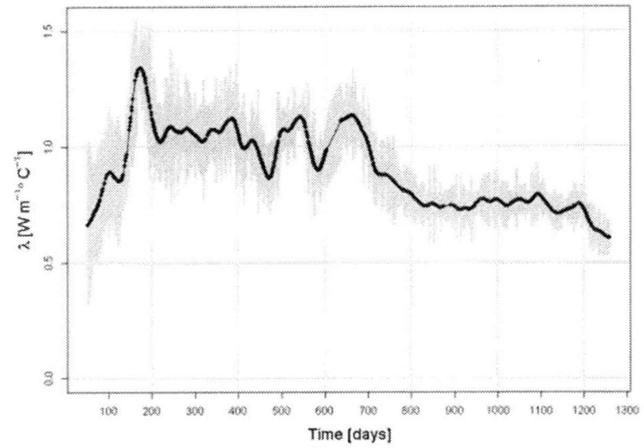

Figure 2. Apparent thermal-conductivity over time (adapted from [6])

An initial verification of this theory was performed by comparing the data analysis results to the experimental analysis of the deposits collected in the refinery. First, the apparent conductivity and thickness were used to estimate the average composition of the deposit by applying the multi-component deposit model in [3] and assuming inorganic and organic to behave as two separate phases composing the deposit. By using a Co-continuous model to describe the deposit as a heterogeneous solid, an average inorganic content of $58_{wt}\%$ was estimated. This result was in excellent agreement with the inorganic content found in the deposits, $52_{wt}\%$.

Conclusions

The Dynamic Data Analysis™ of a refinery heat exchanger allowed calculating the time-evolution of the thermal-conductivity and thickness of the tube-side deposit. This was possible thanks to the availability of pressure drop measurements, in addition to the usual temperatures and flowrates. The analysis helped detect an acute deposition episode in the thickness time profile. The trends observed in the apparent thermal-conductivity indicated that such acute episode was likely due to the fast deposition of inorganic materials, which could be linked somehow to inefficiencies in the operation of the desalter. This interpretation of the results has been supported by comparison with experimental analysis of the deposits.

References

[1] F. Coletti and G. F. Hewitt, *Crude Oil Fouling: Deposit Characterization, Measurements, and Modeling*. Boston: Gulf Professional Publishing, 2014.

[2] F. Coletti and S. Macchietto, "A Dynamic, Distributed Model of Shell-and-Tube Heat Exchangers Undergoing Crude Oil Fouling," *Ind. Eng. Chem. Res.*, vol. 50, no. 8, pp. 4515–4533, Apr. 2011.

[3] E. Diaz-Bejarano, F. Coletti, and S. Macchietto, "A new dynamic model of crude oil fouling deposits and its application to the simulation of fouling-cleaning cycles," *AIChE J.*, vol. 62, no. 1, pp. 90–107, 2016.

[4] E. Diaz-Bejarano, F. Coletti, and S. Macchietto, "Impact of Complex Layering Structures of Organic and Inorganic Foulants on the Thermohydraulic Performance of a Single Heat Exchanger Tube: A Simulation Study," *Ind. Eng. Chem. Res.*, vol. 55, no. 40, pp. 10718–10734, 2016.

[5] E. Diaz-Bejarano, F. Coletti, and S. Macchietto, "Thermo-Hydraulic Analysis of Refinery Heat Exchangers Undergoing Fouling," *AIChE J.*, vol. 63, no. 3, pp. 984–1001, 2017.

[6] E. Diaz-Bejarano, E. Behranvand, F. Coletti, M. R. Mozdianfard, and S. Macchietto, "Organic and Inorganic Fouling in Heat Exchangers – Industrial Case Study: Analysis of Fouling State,"

Appl. Energy, vol. 206, pp. 1250–1266, 2017.
[7] E. Diaz-bejarano, F. Coletti, and S. Macchietto, "Modeling and Prediction of Shell-Side Fouling in Shell-and-Tube Heat Exchangers," *Heat Transf. Eng.*, vol. Accepted f, no. https://doi.org/10.1080/01457632.2018.1446814, 2018.

509071 Improving Catalytic Reformer Heaters Â Duty Using New Technologies

Abstract:

Catalytic Reformer Units are one of the most important units and are always being run at the maximum capacity. The multi stage heater used in the unit is one of the critical equipment. It is supposed to be heating the charge going to the catalytic reforming reactor. The reactor is multi stage and the reaction is endothermic. The charge is again reheated in the first stage, second stage and third stage inter heaters. In most of the units, we find that the charge heater is fired very hard and is generally limited. The inlet temperature to the heater is generally lower than the design leading to unit being limited. Furnace Improvements has developed a split flow process for increasing capacity of these heaters by as much as 30%. We use the convection section to preheat the process instead of generating steam. This increases the heat duty of the heater at a very low cost. We are presenting two case studies illustrating substantial improvement in the heater capacity.

512482 A Kinetic Analysis of the Shipp Synthesis of 2,2',4,4',6,6'-Hexanitrostilbene from 2,4,6-Trinitrotoluene

Abstract:

2,2â€™,4,4â€™,6,6â€™-Hexanitrostilbene (HNS) is an important explosive utilized for multiple purposes due to its thermally stable nature and insensitivity to impact and electrostatic shock. The most common synthesis route used for manufacturing HNS is known as Shipp synthesis and involves using an alkaline sodium hypochlorite solution to facilitate the conversion of 2,4,6-trinitrotoluene (TNT) to HNS. The goal of this study is to develop a reaction network and the corresponding kinetic parameters for the Shipp synthesis method. This reaction network can then be utilized to optimize the reaction conditions and show how the production of HNS using Shipp synthesis could be converted from a batch process to a continuous one. A key issue surrounding this synthesis route involves the low solubility of both the reactants and products, therefore large importance is placed on the solubility modeling of the reacting system.

512964 Halloysite Nanotube Nanocomposite: Processing and Characterizations

Abstract:

Nano-clays have gained increasing popularity in practical applications including functional polymer-clay nanocomposites (PCNâ€™s), material compatibliziers, polymer and rubber processing aids, performance enhancing nanofillers, rheological property modifiers, thermomechanical property enhancers, use in controlled drug release agents, and much more. Among the members that are classifiable under nanoclays, halloysite nanotube (HNT) is of particular interest, based on its abundant availability as high-purity naturally-occurring minerals, unique tubular nanostructure, high aspect ratio, inherently rich functionality, a nearly uniform ratio (2:1) of silica to alumina, excellent cation exchangeability, retainability of both hydrophilic and hydrophobic chemicals, and more. In this paper, fabrication, processing, and characterization of HNT-nanocomposites will be examined and a preferred approach will be presented. The relationship between the thermomechanical and rheological properties and the specifically enhanced functions will also be investigated on the prepared HNT-nanocomposites.

512987 The Development of a Novel Method Using Cryogenic Grinding to Produce Chitin Nanofibers for Blending with Biodegradable Polymers

Abstract:

As plastics continue to buildup in our landfills, the demand for biodegradable polymers is ever increasing as an alternative to conventional polymers, which may take hundreds of years to break down in the natural environment. However, many of the current commercially available biodegradable polymers lack all of the desirable properties of conventional thermoplastic polymers. One such way to improve existing biodegradable polymers in functional applications is by reinforcing them with nanofibers in forms of nanocomposites. Of more recent interest in this area, chitin nanofibers have shown to increase Youngâ €™s modulus and toughness when incorporated into thermoplastic polymers. Chitin is a highly abundant natural biopolymer which is second only to cellulose. Chitin is the primary component of the exoskeletons of most arthropods, including crabs, prawns, and insects. It can also be found in fungi cell walls. Unlike cellulose though, the majority of chitin is just discarded as commercial waste. The exoskeletons are largely made up of a chitin fiber matrix along with minerals and proteins. Once the minerals and proteins are removed, the chitin fibers are strongly held together by hydrogen bonding which can make fibrillation quite difficult. Chitin is also not soluble in most common solvents. There are currently various methods in the literature used to produce chitin nanofibers including electrospinning, grinding with specially designed grinding stones, and high-pressure homogenization, but currently no methods that have been commercialized. Additional research is needed to develop new methods and improve existing methods of chitin fibrillation. In this study, a cryogenic grinder was used to grind commercially purchased chitin with additional modifications made to the grinding process to improve fibrillation. Results were analyzed by SEM, TGA, DSC, and XRD.

518350 Hiring People with Disabilities: Competitive Advantage for the Savvy Manager

Abstract:

John Johnston (PhD, MBA) is the past-chair of the American Chemical Society's Chemists with Disabilities Committee.

John developed the "Scientists with Disabilities" poster series to illustrate the vast potential of persons with disabilities.

These posters have been displayed in numerous government agencies and schools throughout the U.S.

As a means to manage his speech disability, John studied public speaking for many years and has become an award-winning speaker.

John will share an abridged version of his thought-provoking presentation "Hiring People with Disabilities: Competitive Advantage for the Savvy Manager"

with AIChE at the Spring 2018 national meeting.

518356 Navigating the ADA

Abstract:

Sandy is a licensed attorney in both Delaware and Pennsylvania. Throughout her professional career she has had the opportunity to focus on civil rights, and more particularly the ADA people with disabilities. When she first began her career it was a learning experience, and also a great opportunity to use her knowledge and training to make a difference in employment rights for a very under-represented segment of society. Her career began with working as a staff attorney at Delaware Community Legal Aid for the Disabilities Law Program. She knew she had a calling in the area of civil rights after working in the Volunteer Legal Services clinic while in law school.

With the passage of the ADA inn 1990 Sandy began further research, study and direct work for the rights of people with disabilities. As the years have passed, she has also focused on unconscious bias in all areas of civil rights for protected classes. A message she'd like to convey to those she has the pleasure of speaking with is that people with disabilities still have the highest unemployment rate, are entitled to reasonable accommodations in employment, and are subject to the highest rate of bias. We need to move the needle forward and take advantage of the vast talent of people with disabilities. Her focus for her discussion will be on those with disabilities who are capable of bringing much talent, productivity and advancement to employers in this field. The law provides for this. Navigating the ADA can be difficult at times.

She will focus on who is covered by the ADA, all aspects of the employment process, and particularly how to navigate to reasonable accommodation provisions of the ADA.

Within the vast population of people with disabilities, there are individuals with both visible and invisible disabilities. Those with invisible disabilities have formed some of their own support group networks as they seek to obtain gainful employment.

When Sandy was at what would have likely been critical years in her career she was diagnosed with a serious auto-immune disease. So at this point she was advocating for herself and it became more personal. It remains personal based on the relationships she has formed throughout her career.

She transitioned from tradition courtroom law to working in Civil Rights program management and leadership roles in Delaware State Government. Her current title is Statewide Administrator for Equal Employment Opportunity. She has had to opportunity to focus on individuals with disabilities. She was also part of the former Governor of Delaware, Jack Markell's state award winning platform of "Doing the Right Thing: Employment of People with Disabilities". This was Governor Markell's platform when he chaired the National Governor's Association during his tenure.

She continues to be the statewide lead on the American's with Disabilities Act. She regularly works with individuals with disabilities who are seeking employment opportunities. She also spends a tremendous amount of time working on accommodations for employees with disabilities.

Hiring individuals with disabilities is not just the right thing to do. It is also a source of untapped incredible talent from dedicated individuals who want to see their workplace succeed and to contribute to society with their talents and dedication.

During her presentation she will focus on The ADA requirements and the realities that individual's with disabilities often encounter. She believes that the reasonable accommodations guaranteed through the ADA should be discussed and further explained.

18 AIChE Spring Meeting
14TH GLOBAL CONGRESS ON PROCESS SAFETY
ORLANDO FLORIDA 2018

"Unmanned Aircraft Systems (UAS): Case Study that Highlights Challenges and Opportunities"

Robert Shirley, Manager
Premium Inspection & Testing
Capitol Division
rshirley@pitinc.com

Damien Parson, Fixed Equipment Manager
Florine Vincik, Process Safety Expert Services
BASF
Geismar, LA 70734
damien.parson@basf.com, florine.vincik@basf.com

Prepared for Presentation at
American Institute of Chemical Engineers
2018 Spring Meeting and 14th Global Congress on Process Safety
Orlando, Florida
April 22 – 25, 2018

"Unmanned Aircraft Systems (UAS): Case Study that Highlights Challenges and Opportunities"

Robert Shirley, Manager
Premium Inspection & Testing
Capitol Division
rshirley@pitinc.com

Damien Parson, Fixed Equipment Manager
Florine Vincik, Process Safety Expert Services
BASF
Geismar, LA 70734
damien.parson@basf.com, florine.vincik@basf.com

Keywords: Emerging Technologies, Unmanned Aircraft Systems, Drones, Inspections

Abstract

This case study provides insights into the use of unmanned aircraft systems (UAS) or drones in an industrial setting. Challenges are significant with respect to regulatory entities such as Federal Aviation Administration (FAA) and Department of Homeland Security (DHS) as well as on-site risk assessment and coordination. Opportunities for cost avoidance, reduced resource usage, improved safety, and downtime minimization are benefits of using these systems for inspections in industrial assets.

1 Background

During a Louisiana Chemical Association LCA conference a Geismar, LA-area Plant Manager delivered a presentation that stressed the need for industry to move towards Smart Manufacturing by embracing new technologies, such as drone use, 3D printing and robotics, to stay competitive, while increasing safety and efficiency. A local company, Premium Inspection & Testing, met with BASF team leaders and explained the benefits of drone usage for mechanical inspections at the facility. A follow-up meeting was scheduled to determine visual inspection capabilities for a problematic flare stack. An Unmanned Aircraft System (UAS) team was quickly mobilized and delivered unbiased, real-time data on the asset in a matter of a couple of hours. The asset owner and team were impressed with the safety procedures, live feed capabilities and the quality of data, that enabled them to make a decision that evening. What would have traditionally required cranes, man lifts and scaffolding, ended up saving resource time and money, while most importantly improving safety measures. By utilization of the UAS technology, other less safe and costly activities such as cranes, man lifts and rope access were avoided. This technology has led to the increased demand for inspection of confined spaces and tanks.

Safety of personnel in operating industrial facilities is always a top priority. Elevated or confined space process structures, as with flare stacks, cooling towers, and columns/reactors, pose significant risks to personnel during the inspection process. Risks for industry were noted by the 2016 OSHA Call to Action to preventing fatalities from falls.

This UAS case study prompted the need for developing global and regional programs to address UAS missions and hazards to people and assets. Significant time (months) was needed to establish these programs.

2 Case Study: Challenges to Introduction and Use of UAS at an Industrial Facility

2.1 Corporate Guidance

The use of unmanned aircraft systems (UAS) require a high-level of due diligence regarding site safety and security. It is recommended that a UAS Operator has a Safety Management System (SMS) in place that documents and evidences an organized approach to managing safety, including the necessary organizational structures, accountabilities, policies and procedures. An organizational risk assessment and management process (including a risk register) should be implemented and maintained for all UAS operations.

Following established process safety practices such as management of change, standard operating procedures, and risk assessment may not cover the full extent of drone usage. Use of drones must also comply with federal, state, regional and local regulations and requirements. Therefore, corporate guidance as well as a specialized approval team is highly recommended to provide a consistent approach for drone usage throughout a company that is globally based. Corporate level responsibility includes keeping a finger on the pulse of developing rules and regulations and communicating corporate rules down to the regional level. The regional and local levels must then form a multidisciplinary team of operations, engineering, purchasing, safety and security personnel to evaluate drone usage policies/procedures for specific requests at individual locations.

Site permission and flight planning in the form of a *plan of activities* should be developed by maintenance engineers familiar with the use of UAS. It is recommended that prior to any UAS operations, the UAS Operator (internal or third party) should submit a plan of activities to the asset owner in accordance with any established procedures relevant to the flight area. The plan of activities should include at least the following:

1. Nature/objectives of the flights
2. Dates and times for all flights
3. Name and contact details for the UAS operations scheduling and management.
4. Name and contact details for the UAS operations flight team (e.g. pilot-in-command and camera/payload operator)
5. The product details and serial number of the UAS device to be used
6. Site visit/inspection reporting (recommended where possible)
7. Maps or diagrams of the flight area
8. Infrastructure subject to inspection
9. Planned altitudes for the UAS operation
10. Planned take-off, recovery and return-to-home locations.

Emergency scenarios and procedures must be developed for:

- Loss of control requiring redundancies
- Collision and restricted areas
- Mechanical or electrical failure
- Loss of line of sight requiring return to home safeguard
- Sudden changes to environmental conditions
- On site emergency situations (e.g. access for emergency services, fire and site evacuation).

An on-site work permit may be required (and is different from permission to fly). It is the responsibility of the asset owner to assess whether a work permit is needed. All work permits and results of risk assessment should be approved by the asset owner. If the asset is hosted at a third-party location (such as pipeline inspections), the third-party location owner's authorized representatives may also be required to grant an on-site work permit. The work permit should be valid for each individual UAS operation and for the maximum duration needed to complete the inspection work (usually less than twelve hours).

It is recommended that a UAS Operator has secured third-party public liability insurance, professional indemnity insurance and employers liability insurance to insure against the risk of any significant hazardous outcomes (i.e. personnel health and safety and property damage) resulting from UAS operations.

2.2 Regulating Agencies

A Part 107 UAS License to operate a UAS is required in an industrial setting. Both the Federal Aviation Administration (FAA) and the United States Coast Guard (USCG) require 48 hours notification before any scheduled drone flight [1]. UAS Facility Maps show the maximum altitudes around airports where the FAA may authorize part 107 UAS operations without additional safety analysis. The maps should be used to inform requests for part 107 airspace authorizations and waivers in controlled airspace.

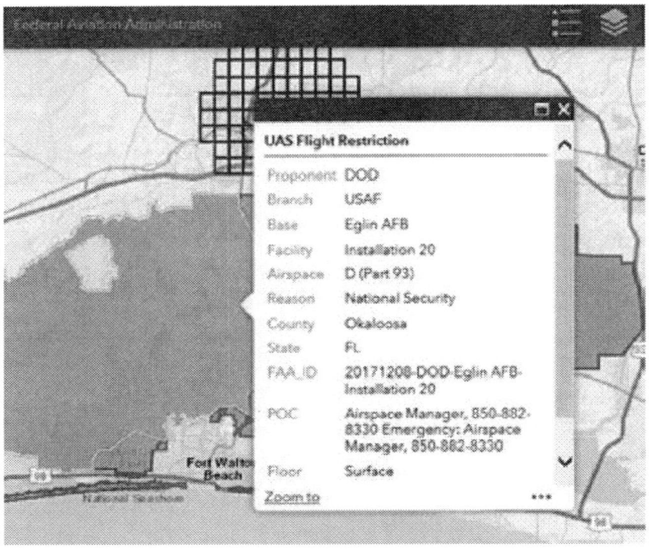

Figure 1: FAA Map Showing UAS Flight Restriction (shown in red solid coloring)

These maps DO NOT authorize operations in these areas at the depicted altitudes – they are for informational purposes ONLY. Operators must still apply to operate in controlled airspace (Class B, C, D, or surface area E) by completing a Part 107 Airspace Authorization or Waiver Application in the *FAA Drone Zone* section of the website [1].

Individuals who request a part 107 airspace waiver and authorization are encouraged to consult the maps prior to submitting a request to determine locations and altitudes that can be approved quickly. [1]

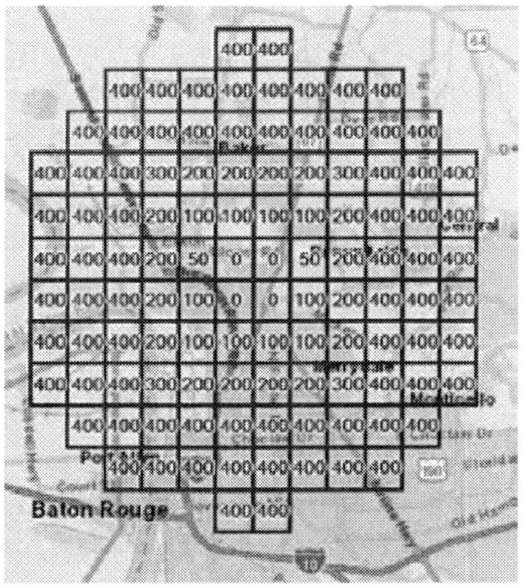

Figure 2: FAA Map Showing UAS Altitude Restrictions in Feet

In this case study, it was quickly determined that a pre-flight checklist was needed to incorporate this FAA flight waiver/airspace authorization request online with the map altitude to operate in a specific location. Drones that are under 55 pounds are considered "Small Unmanned Aircraft Systems (UAS)." Authorization requests for a larger drone or altitude more than the map altitude does require further coordination [2]. An example of the use of these is for pipeline construction that can handle payloads up to 500 pounds. Other emerging markets is offshore supply transport to oil rigs with authorization by the US Coast Guard.

2.3 Lease, Own or 3rd Party

Most drone manufacturers do not provide a retirement life cycle. There is no industry standard for when to take your drone out of service. Through experience, this case study contractor has settled upon 500 hours (actual) flight time for their hexa-copter, and the smaller confined space drone is closer to ~ 300 hrs. This lends well to lease vs purchasing a drone, assuming the utilization is there.

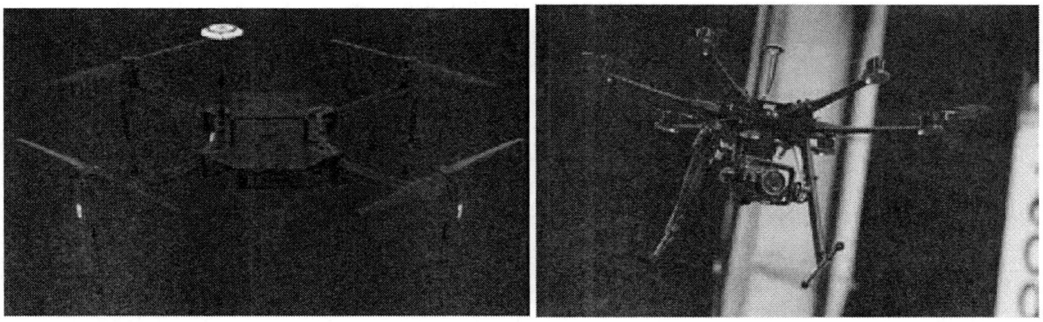

Figure 3. Quad and Hexa Copter UAS (photo courtesy of DJI)

Anyone can go out and get a Part 107 UAS license. It's like acquiring a driver's license. After passing a test, one may have to fly their drone, but overall the process is relatively easy for someone to pass. However, that doesn't mean the individual is competent to fly a drone inside a chemical, petrochemical, or manufacturing facility. Flying a drone is not like riding a bike. It takes consistent and repeated flight time to keep up this skill. In this case study, nested pilots are estimated to spend at least 50% of their work week (20 hrs.) performing drone work (paperwork, notifications, mock flights, scheduled flights, drone maintenance, etc.). Regardless of utilization time, practice is needed to ensure drone use skill.

This case study company utilized a company that had the most flight hours of any drone service provider in the local area, and with that experience came knowledge of what to expect, what to do when something goes wrong, and how to not place the drone in a compromising situation. Their safety mindset and protocols are second to none, including employing licensed commercial pilots and implementing that experience into their UAS protocols. The future strategy from this case study is to have a UAS contractor train a couple of selected inspectors for 6-8 weeks to allow them to perform inspections on site. Additionally, it is a best practice that UAS pilots maintain training and documentation of their flight times and possess minimum flight experience prior to performing the duties of a UAS pilot-in-command (PIC) [3]. When operating in an industrial environment, the following flight currency should be met to ensure pilots are properly qualified.

1) 50 hours flight time in the preceding 120 days performing the duties of a UAS PIC; and
2) 20 hours flight time in the preceding 60 days performing the duties of a UAS PIC in the make and model of UAS to be flown.

It is also recommended that the following types of checklists be reviewed, completed and verified pre-flight [1]:

- Location/on-site inspection
- UAS device and equipment inspection and test
- Ground control systems inspection and test
- Pilot/camera operator readiness

2.4 Risk Assessment

Due to the inherent risks of UAS operations, it is recommended that a risk assessment face to face meeting is completed and documented before a UAS operation, and should involve drone user, operations, maintenance, process safety, occupational safety, inspection data end-user responsible for the use and acceptance of any data. This multidisciplinary approach is essential in evaluating hazard scenarios, drone trouble consequences and failure modes, electrical classification of equipment to be inspected, and safeguards for prevention, protection, and mitigation. The following techniques are useful:

- Failure Modes and Effects Analysis (FMEA), What-If Hazard Method,
- Job Safety Analysis (JSA) and
- Work Permit

Multiple risk assessments may be required for the same flight area on the same day (e.g. due to weather conditions, changing inspection scope). The use of pre-determined risk assessments designed to cover generic and/or historical conditions and variables is not allowed. Customized and specific risk assessments are considered industry best practice as each risk situation should be assessed on its own merits, and use of a generic assessment to cover all potential issues can lead to complacency amongst personnel [3].

The risk assessment should typically address the following criteria:

2.4.1 Flight stability and accuracy risks:

- Current weather
- Predicted weather during UAS operations

- Environments with excessive hot or cold temperatures
- Air flow from ventilation fans, both intake and exhaust
- GPS denied areas
    - Confined spaces have GPS denied environments. These include tanks, vessels, pipes, boilers, reactors, cogens, silos, refractory lines, etc. Since airspace is blocked from the roofs it does not allow for a drone equipped with GPS to connect to the satellites and operate in that environment.

2.4.2 Drops risks mitigated by Communication

- Flying over areas occupied by personnel
- Flying over areas potentially occupied by personnel
- Flying over structurally vulnerable equipment or assets

2.4.3 Fire and Explosion risks mitigated with Fire Protection Systems

- Battery fire
- Running or rotating machinery.
- Areas where explosive gases or materials may be present.

2.4.4 Collision risks mitigated with Line of Sight Co-Pilot

- Birds (Clearing area prior with Noise Cannons)
- Unguarded ventilation fans
- Other UAS
- Piloted aircraft (e.g. supply helicopters)
- Areas where line of sight cannot be maintained
- Confined or limited access spaces
- Height/flight ceilings
- Areas with falling debris or water
- Asset structure
- Other vessels or vehicles (e.g. powered equipment)

2.4.5 UAS device communication drop-out risks:

- Radio or signal transmitters that could affect communication with the UAS device.
- High voltage cables or areas that may affect control signals or damage the UAS device.
- Dense areas of structural steel.

2.4.6 Is the area, infrastructure or asset being inspected in a “live” or “downtime” status.

2.4.7 Any known/discovered design or operational limits of the UAS device.

The following risk assessment principles and methods in IEC/ISO 31000 Risk Management Principles and Guidelines or ISO/IEC 31010 Risk Management - Risk Assessment Techniques were used in this case study.

A risk assessment may last a full work day prior to allowing a typical mission which may consist of only 1.5 hours in an established industrial site:

- Work Permit/Risk Assessment/Setup 30 mins,
- Fly Mission 30 mins,
- Recover/Close Permit 30 min.

Challenges with on-site communication between third party maintenance and engineering and approvals to work have created delays in missions for the UAS vendor. Defining acceptable conditions for work and stop work authority were critical outputs in the risk assessment process. Having a site-trained person in attendance was critical for effective understanding of production non-steady state operating activities upsets or transition periods and site alarms.

After corporate and site procedure were developed, the management of change process was not used unless deviations required risk assessment and approval. An example of where a deviation might occur is in use of a non-approved, third-party contractor not listed in the contractor safety database. An assessment of UAS contractor safety records is an important aspect in industrial applications.

Regarding drone type, an assessment of failure mode and effect analysis (FMEA) is an excellent technique that was a deciding factor in this case study to utilize a hexa-copter (6 motors) vs a quad copter (4 motors) shown in Figure 3. If one of the motors on a hexa-copter fails in flight, the drone would still fly and land safely. If one of the motors on a quad copter fails, … crash … the question is, crash into what? In the case of a vessel entry into a confined space, the ball drone shown in Figure 4 was chosen due to its gyroscope technology enclosed in a carbon fiber protective cage that enables it to maneuver and bump into walls and equipment without damage. All accidents and near-miss incidents should be documented and reported to the asset owner representatives. A closeout meeting is recommended to review the results and discuss inefficiencies, incidents.

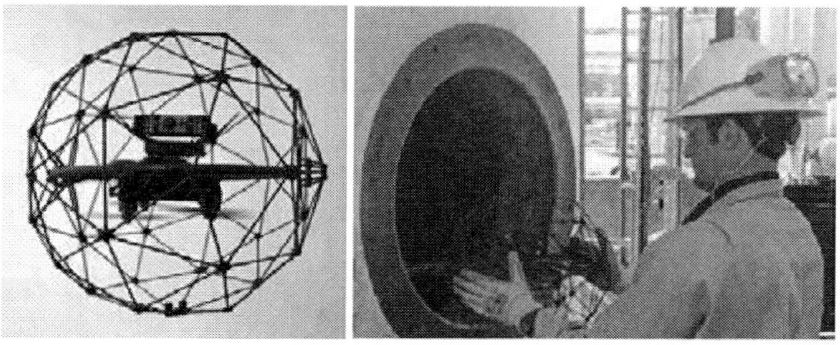

Figure 4. Ball-Type Drone with Gyroscope and Carbon Cage (Courtesy: LandBros)

Figure 5. Case Study Confined Space Mission Pictures (Courtesy: LandBros)

Drones are limited based on battery life, typically 25 minutes (slightly less for an internal inspection drone with a cage). However, drone operators keep many extra batteries which can be quickly changed. When the battery is extremely low, there is an automatic return feature to prevent losses. Transponders with unique identification codes are also useful for tracking drones beyond the line of sight (BLS). GPS positioning and weather does have an impact on drone external and internal inspections.

Security is notified and they make notifications to area facilities as well as a site-wide announcement to alert people of a drone usage. This is to ensure people are clear of the general area (falling object risk) and to alert personnel that the risk to critical infrastructure has been evaluated to deter threat reports. In the initial use of drones on site, notifications made during a drone pre-flight checklist did initially draw large crowds simply out of curiosity of this emerging technology.

3 Case Study: Benefits from Use of UAS at an Industrial Facility

Use of an emerging technology, unmanned aircraft systems or UAS (or drones) provided this case study unbiased, real-time data on equipment captured many angles not typically captured by crane or man lift access. Use of a third-party UAS company provided a comparative cost estimate of Non- Destructive Technique (NDT) inspection capabilities for business case development. In this case study, several drone inspections were done and primary benefits were (1) the ability to see the data immediately and collaboration with on-site inspection department personnel to concentrate videos and pictures in trouble spots real-time, and (2) dual feeds for both high resolution and thermal imagery that allowed the inspectors to quickly provide an inspection report including video and picture files on DVD to the inspection department engineer who then could communicate quickly with operations to make a decision regarding future work. Streaming of real-time video to remote locations are available but clients must provide internally-developed encoders to transmit the signal on a secure network with ethernet or hotspot.

Table 1: Case Study Resource Efficiencies

Flare Inspection (unplanned)	60% resource savings
Flare Inspections (2)	80% resource savings
Electrical Substation Thermal Imagery	90% resource savings
Confined Space Inspection	80% resource savings
Unit Transformer Inspection	90% resource savings
Two-dimensional top-down photos of units and oblique angles	80% resource savings

Figure 6. Case Study Confined Space and Flare Mission Pictures

Typical applications in this case study included visual inspections and thermal inspections in difficult to access areas. Drones were also used for corrosion under insulation detection, detection of electrical "hot spots", flare inspections, pipe rack inspections, and wire tray inspections. A few other beneficial examples of the use of drones in industry include aerial photography, pipeline external inspections, fence line security monitoring, and damage assessment. Regarding cost, UAS inspections in this case study have realized 70-80% cost savings compared to traditional methods of inspection.

4 Conclusion

According to the Department of Homeland Security, since UAS use in the United States has increased as a cost-effective, versatile business and national security tool, as well as a popular recreational hobby, the Federal Aviation Administration (FAA) estimates combined hobbyist and commercial UAS sales will rise from 2.5 million in 2016 to 7 million by 2020 [2]. Therefore, an industrial facility must understand the threats to critical infrastructure and remain vigilant to updated regulations and due diligence with notifications of UAS activities to regulating agencies.

Utilizing drones in this industrial case study was a complex task to implement at first until familiar risk-based, holistic approaches already used on site were used to ensure compliance with regulations and zero threat of exposure to people and infrastructures both on site and off site. Several benefits realized in an industrial facility (depending on the job) may include:

- Eliminate the need for people to access elevated and confined work spaces,
- Save time and resources in preparation of equipment for inspection,
- Reduce operations cost by reducing downtimes and inspection costs,
- Provides photos and videos for a permanent record of equipment inspection.

Future applications of autonomous drone missions will be in the areas of mechanical work (robotics), security inspections, and routine inspections/thermal imagery with futuristic *artificial intelligence* to constantly take data on asset capabilities and provide feedback for plant personnel use. It's not science fiction any more…it's science fact!

5 References

Web resources:

[1] Federal Aviation Administration Available at https://www.faa.gov/uas/request_waiver/uas_facility_maps/ Accessed on February 3, 2018.

[2] Department of Homeland Security Available at https://www.dhs.gov/publication/uas-fact-sheets Accessed on February 3, 2018

[3] "Guidance Notes for Inspection Using Unmanned Aircraft Systems", Lloyd's Register, March 2016

Composite Intermediate Bulk Containers (CIBCs): Understanding Risks & Emerging Fire Code Issues

Michael Snyder, PE CSP CFPS
DEKRA Process Safety
113 Campus Drive
Princeton, NJ 08540
Mike.Snyder@Dekra.com

Prepared for Presentation at
American Institute of Chemical Engineers
2018 Spring Meeting and 14th Global Congress on Process Safety
Orlando, Florida
April 22 – 25, 2018

Composite Intermediate Bulk Containers (CIBCs): Understanding Risks & Emerging Fire Code Issues

Michael Snyder, PE CSP CFPS
DEKRA Process Safety
113 Campus Drive
Princeton, NJ 08540
Mike.Snyder@dekra.com

Keywords: NFPA 30, OSHA 1910.106, Large loss fire events, Combustible liquid fire risk, Plastic containers, Pool fires, Composite intermediate bulk container

Abstract

The use of Composite Intermediate Bulk Containers (CIBCs) for the storage and handling of Combustible Liquids has introduced significant fire risks into storage, handling, and transportation operations. Beginning with the sentinel research work conducted by the UK Health & Safety Executive in the mid-2000s, continued research and educational efforts have resulted in improved understanding of the fire risk controls required to manage CIBC hazards. Several large loss fires involving CIBCs continue to demonstrate remaining challenges with adequately managing this hazard.

NFPA 30 (Flammable and Combustible Liquids Code) serves as the foundation for flammable and combustible liquid regulation in the United States. New requirements to manage CIBC fire risks have been established in NFPA 30 (2018) requiring the use of listed and labeled CIBCs with Class II and IIIA liquids. Simple evaluation flowcharts are provided to identify allowable storage configurations that are permitted by NFPA 30 (2018) for CIBC storage.

Historically, the management of pool fire risks presented by rapid failure of CIBCs in fire scenarios has focused on the design of containers to resist the heat exposure from fire, ultimately preventing catastrophic failure. New approaches to drainage system design have entered the market and insurance company approval standards, which provide additional possibilities for limiting pool fire size. Ultimately, this may provide an additional method for fire risk management of CIBCs.

1 Introduction

Composite IBCs (CIBCs) have become a popular container for the transportation and storage of combustible liquids. The typical CIBC consists of a blow molded polyethylene bottle that is placed inside a wire cage or sheet metal enclosure. While authorized in sizes ranging from 119 gallons (450 liters) to 793 gallons (3000 liters), the most commonly used container is sized at 1 cubic meter (275 gallons; 1000 liters)

The widespread popularity of the container is attributed to the significant storage and handling efficiencies they provide. Improved utilization of warehouse and transportation spaces occurs, since a 275 gallon (1000 liter) CIBC utilizes the same footprint of a pallet containing four 55-gallon (208 liter) drums -- providing a 25 percent nominal increase in storage and logistics efficiencies. Composite IBCs also provide filling and dispensing efficiency, allowing one action to be established for the transfer of up to 275 gallons (1000 liters) versus the need to run multiple operations to accommodate five 55- gallon (208 liter) drums.

Figure 1: Typical CIBCs Storing Combustible Liquids (Image provide by the Fire Protection Research Foundation and Property Insurance Research Group)

2 Exploring the Pool Fire Risk Presented by CIBCs

The main fire protection concern is that CIBCs exposed to fire fail quickly, thus creating large pool fires by adding a significant amount of fuel to an existing fire. The large fuel release can overwhelm a building's fire protection and loss control features.

The UK Health and Safety Executive conducted sentinel research [1] on this hazard in the early 2000s, following a series of large loss fires in Europe involving CIBCs. The results of this research documented:

- The initial fire exposure to CIBCs are typically external to the container, from either an adjacent use of flammable liquids or ordinary combustible materials. Many CIBCs are vulnerable to rapid container failure from small exposure fires involving ordinary combustible materials (e.g. wood and paper).
- Upon exposure to elevated temperature from fire exposure, the thin blow molded polyethylene bottle allows extensive permeation of the container contents to occur. This effect is significantly more pronounced with hydrocarbon fluids than with polar solvents.
- The permeation softens the plastic to the point that within 2 to 3 minutes of fire exposure, there are a number of holes created in the CIBC, and the container contents are rapidly lost.
- When the contents of the failed CIBC are combustible liquid, they will ignite and create a large pool fire that exposes additional containers.
- The cascading effect of failing CIBCs overwhelms the design basis of most fire protection and drainage systems resulting in significant fire lost.

A simple analysis of the heat release rate and pool fire size that can be generated by the catastrophic failure of just one CIBC of a typical lubrication oil is summarized in Figure 2 below. The size of the pool fire and heat release rate has driven the historic focus on CIBC fire protection strategies on the development of fire resistant containers that prevent leakage from CIBCs exposed to fire. In the model codes and standards, these "fire resistant" CIBCs are defined as "Listed and Labelled" containers. In the US, Listed and Labelled containers typically represent CIBCs that have successfully passed testing in accordance with either the UL 2368 or FM 6020 testing standards.

3 Large Loss Fires Involving CIBCs

Large loss fires involving CIBCs occur with some frequency in industry. In all of the reported cases, the basic failure mechanisms identified in the sentinel UK Health & Safety Executive testing have been demonstrated to occur consistently and predictably. A small list of major fire events is provided to highlight the predictable and reproducible dynamic of a large loss fire involving CIBCs. The author is also aware of several significant large loss fires, where the significance of CIBCs was overlooked by investigation and inspection personnel.

Distillex

This fire occurred in April 2002 in North Shields, England at a solvent recovery firm [2]. Sparks from an angle grinder ignited CIBCs stacked three high. Large pool fires occurred from rapid failure of the CIBCs, exposing unvented steel drums. Rocketing steel drums resulted in significant offsite effects, and limited on-scene firefighting activities. Significant environmental damage occurred on site from the uncontained pool fires.

Depth of spill	0.125 in.	0.25 in.	0.5 in.	1 in.
Area of spill	3,529 sq.ft.	1,765 sq.ft.	882 sq.ft.	441 sq.ft.
Heat Release	557,480 BTU/sec	278,808 BTU/sec	139,221 BTU/sec	69,221 BTU/sec
	588 MW	294 MW	147 MW	73 MW
Duration	1.03 min	2.06 min	4.13 min	8.25 min
Flame Height*	69 ft	54 ft	43 ft	34 ft
# of Package Boilers**	803 boilers	401 boilers	200 boilers	100 boilers

* *Method of Thomas - Reference: SFPE Handbook of Fire Protection Engineering, 2nd Edition, 1995, Page 3-204.*

** *Assuming 2,500,000 BTU/hr Boiler*

Typical Package Boiler		
83 BTU/sec	to	694 BTU/sec
300,000 BTU/h	to	2,500,000 BTU/h

Figure 2: Typical Heat Release Rates for Uncontained Pool Fire Arising from Catastrophic Failure of 275 Gallon CIBC Containing Hydrocarbon Lube Oil.

Magnablend

The incident occurred in October 2011 at a liquid blending and storage facility in Texas. The fire began in a blending portion of the facility from an unexpected reactive chemicals event. The fire extended into a warehouse holding several hundred CIBCs containing a variety of hydrocarbon based liquids. The fire quickly progressed through the storage area, creating a large number of rapidly failing CIBCs. The large pool fire overwhelmed fire protection and drainage features of the building.

As the drainage was overwhelmed, a fire department ladder truck was exposed to burning liquid and ultimately destroyed. Fortunately, no firefighters or facility employees were significantly injured. The facility was a total loss. This fire served as the basis for the NFPA case study entitled "The Fire Risk of Composite Intermediate Bulk Containers" [3]

Chemie-Pack

A fire started in January 2011 at a chemical packaging and distribution center in the Netherlands, when an open flame was used to resolve a frozen transfer pumping system handling flammable liquids. The fire grew and exposed approximately 250 CIBCs that were stored in the facility. The CIBCs rapidly failed created a massive pool fire that proved difficult to extinguish. The fire took nearly 24 hours to control. The pool fire resulted in significant environmental damage and the bankruptcy of the company.

A thorough investigation from Dutch Safety Board was conducted, and subsequent prosecution of responsible management personnel ensued [4]

North American Manufacturing Facility (April 2015)

Sparks from a hot work operation ignited a single CIBC containing hydrocarbon-based lubrication oil (Flash Point 142 F/61 C) in a maintenance shop, which was protected by automatic sprinklers. Following several minutes of fire exposure, the container failed and discharged its burning contents to the floor.

Through the duration of the fire, 91 sprinkler heads activated, and were able to be supplied by the facility's superior fire protection infrastructure that allowed spare online fire pumps to be activated. In a typical manufacturing structure, the fire area of >8,000 square feet would have overwhelmed the fire sprinkler system. Even with the superior fire protection, damage in excess of $1 Million USD occurred. [5]

4 NFPA 30: A Basis for Flammable & Combustible Liquid Regulation

Since 1913, NFPA 30 has served as a leading source of practices for the safe handling and storage of flammable and combustible liquids. From 1913 to 1957 document was written as a model municipal ordinance known as the *Suggested Ordinance for the Storage, Handling, and Use of Flammable Liquids*. In 1957, the format was changed to a code, although the technical requirements and provisions remained the same. Since its inception, numerous revised editions have been published as dictated by experience and advances in technology.

NFPA 30 is generally considered a Recognized and Generally Accepted Good Engineering Practice (RAGAGEP) in the field of Flammable & Combustible Liquid Risk Management, and uses a fully consensus oriented process to vet and review updates to the code during each revision cycle.

Composite IBCs were originally recognized as an authorized container in the 2003 edition of NFPA 30, and over time specific requirements and restrictions for their storage and use were developed based on an extensive group of fire tests involving CIBCs [6]. Up through NFPA 30 (2015), the following requirements have existed for CIBCs:

- They are not authorized for the storage of use of Class I (flammable) liquids, except in specific, limited, situations that are detailed in Section 9.1.4 of the Code.
- If CIBCs that are not listed and labelled (passing a recognized fire test) are used, the storage configurations are limited and classified as Unprotected Storage
- If listed and labelled CIBCs are used, stored in approved configurations, and provided with adequate sprinkler protection, an unlimited quantity may be stored in specified palletized arrays and rack configurations.
- Specific allowances for a limited number of CIBCs containing Class II and III Combustible Liquid are permitted for "operational areas".

Many state and regionally adopted Flammable and Combustible Liquid Codes are based on recent editions of NFPA 30. In these jurisdictions, the requirements for CIBC use and storage have the force of law. The International Fire Code (IFC) makes specific references to the approved container lists from NFPA 30 (Section 9.4), so in jurisdictions that adopt the IFC, this aspect of NFPA 30 is also enforced. OSHA 1910.106 is based on NFPA 30 (1969), and does not formally recognized CIBCs as an authorized container. Specific letters of interpretation and agency directives allow the use of these polyethylene containers if they are DOT/UN authorized. [7]. Since OSHA relies on such a dated version of NFPA 30, the reality is that most local and state Fire Codes & Flammable Liquid Codes create more stringent requirements to a facility than the OSHA standard does.

5 NFPA 30 (2018): New Requirements for CIBC Storage & Use

As part of the continued focus on addressing the pool fire risks presented by CIBCs, the 2018 edition of NFPA 30 [8] confirmed and clarified several critical safety aspects of the code, along with establishing two new requirements for the storage and use of CIBCs:

Existing requirements:

- They are not authorized for the storage of use of Class I (flammable) liquids, except in specific, limited situations that are detailed in Section 9.1.4 of NFPA 30 (2018)
- The concept of Protected and Unprotected storage is maintained, with additional guidance provided in the appendix. The limitations on storage quantities and configurations for Unprotected Storage were maintained.

New Requirements:

- Any CIBC used for Class II and IIIA combustible liquids must be Listed and Labelled. CIBCs that are not Listed and Labelled are prohibited for use in this service.
    - To obtain the allowances for Protected Storage, CIBCs must be stored in approved configurations with required minimum sprinkler protection densities.
- Significant changes were made to Section 18.5.4.1 related to allowances for storage of liquids in Incident Use applications. New requirements for conducting a Hazard Analysis and implementation of a new (lower) threshold for Maximum Allowable Quantities of liquids has been implemented. This will reduce the number of CIBCs that can be maintained in an Operations area.

To help simplify the complex decision-making process for appropriate CIBS selection for a given liquid service, and how to determine allowable storage configuration in NFPA 30 (2018), two flow charts have been developed. The charts are provided, below, in Figures 3 and 4.

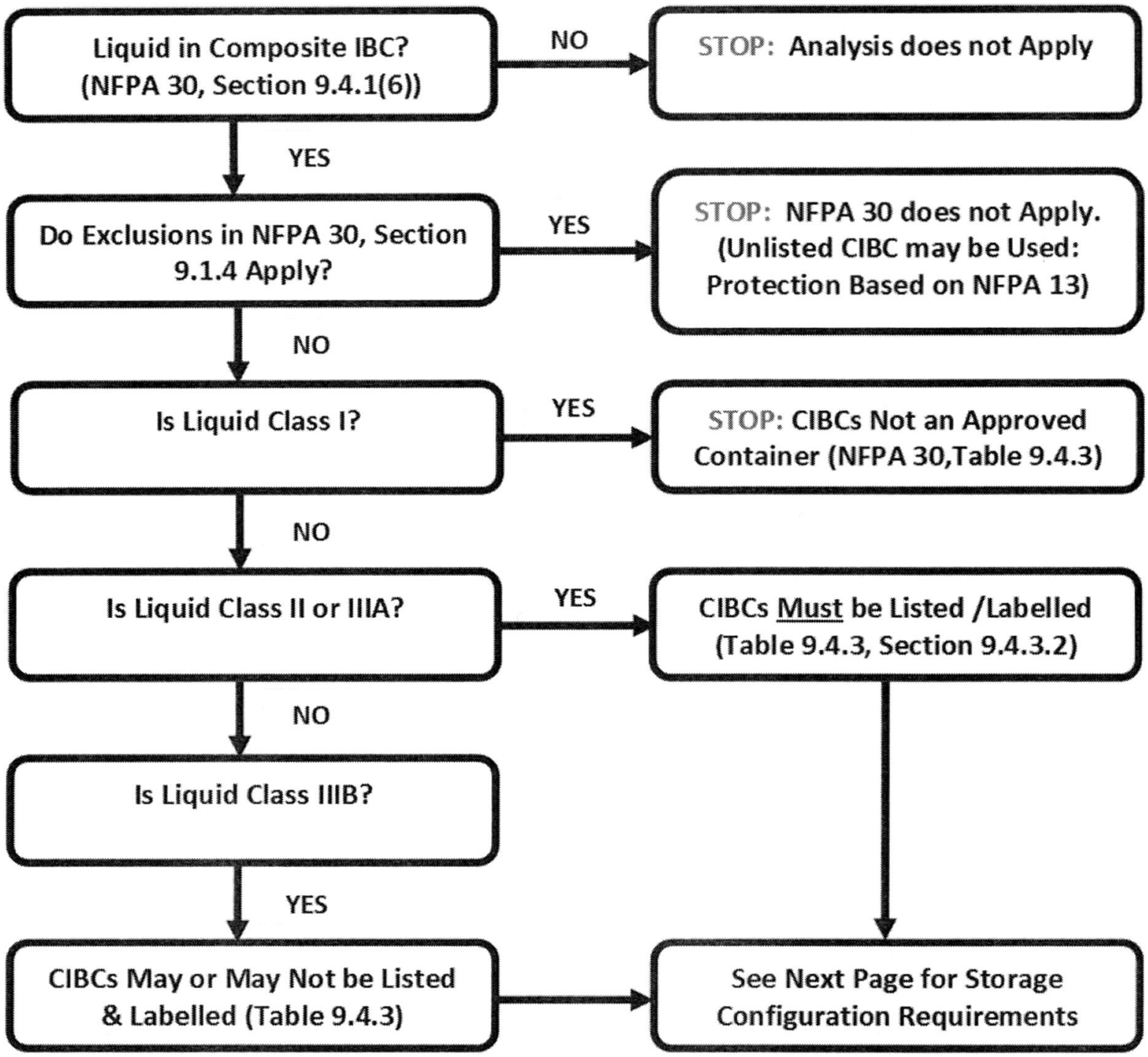

Figure 3: Approved CIBCs for Various Liquid Services based on NFPA 30 (2018) Requirements

6 Emerging Developments in Pool Fire Risk Management

The vast majority of research into the CIBC fire risk situation has been focused on how to enhance the fire and heat resistance of the CIBC, with the objective to prevent creation of large pool fires introduced by container failure. The current approach is best documented in the two US listing & labelling standards for CIBCs, namely UL 2368 and FM 6020.

Recently, a novel approach to CIBC fire risk management has been proposed, which involves the use of an "active drainage system". Essentially, this approach focuses on limiting the pool fire size by channeling a spill fire and directing it to containment (thereby limiting pool fire surface area and heat release rate), along with the active application of cooling water into the drainage

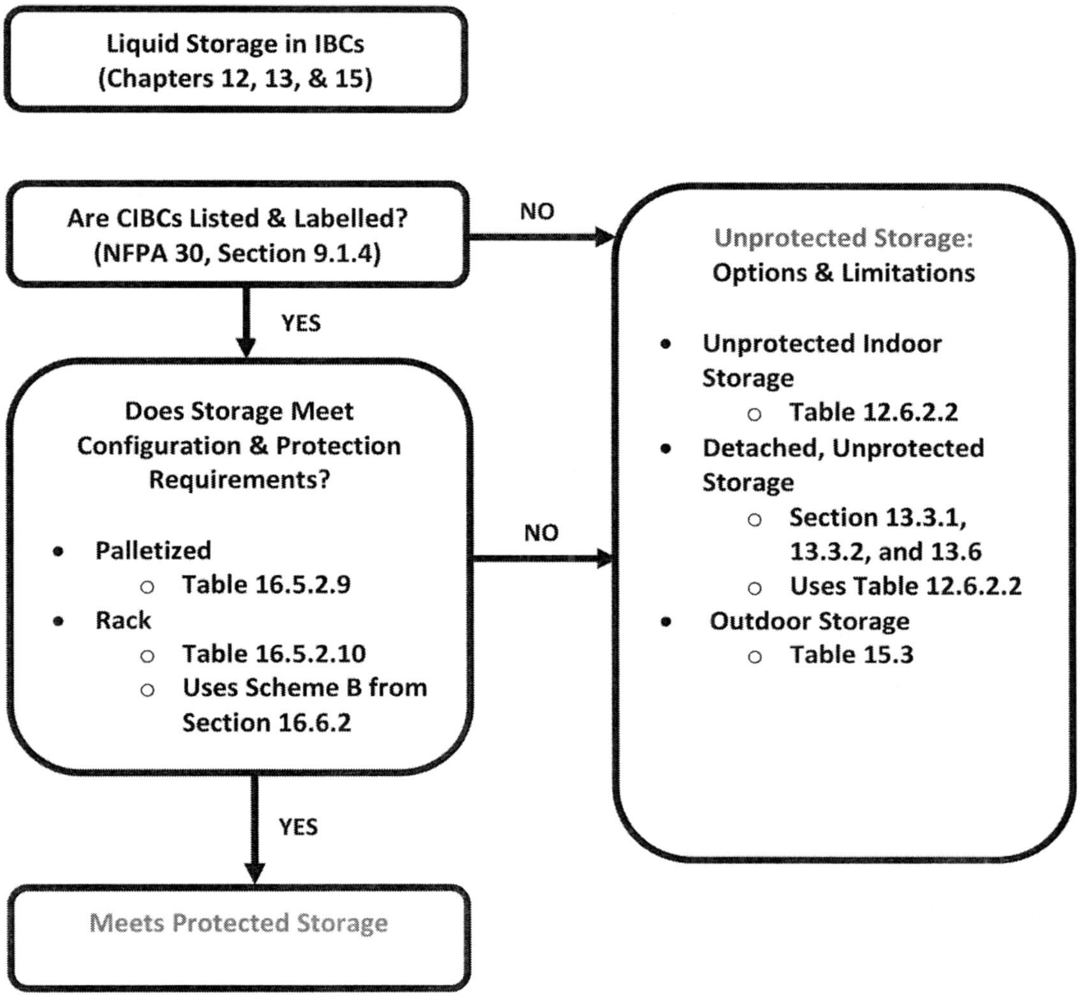

Figure 4: Determining Allowable Protected and Unprotected Storage Configurations for CIBCs in Combustible Liquid Service

system. If effective, this approach can reduce the development of a spreading combustible liquid pool fire from a composite IBC fire event, thereby reducing heat release rate and fire size. The result would be to manage the needed sprinkler design for a storage or operations occupancy handling CIBCs.

Recently, FM Global introduced a new Approval Standard 6090, Ignitable Liquid Drainage Floor Assemblies [9], which focuses on testing and approval of products using this approach. It is desired to discuss this type of loss control approach for the next revision of NFPA 30.

7 Empty CIBCs Present Additional Fire Risks

While the main purpose of this paper is to focus on the hazards of large pool fires that can be created during fire exposure to unlisted CIBCs, the fire risks presented by empty CIBCs must also be considered. Recent fire tests conducted by FM Global have highlighted that stacked arrays of empty CIBCs present a significant fire load and heat release rate, similar to storage of exposed, unexpanded, plastic commodities. [10]

Protected facilities will require a robust sprinkler protection design to adequately control a fire from palletized storage, 3 high, of CIBCs. The FM Global report details recommended protection schemes based on the type of empty CIBC being stored.

8 Conclusion

Fire protection research, and industrial fire loss experience, continues to highlight the hazards presented by the storage of combustible liquids in CIBCs. NFPA 30 has served as a Recognized and Generally Accepted Good Engineering Practice (RAGAGEP), which provides a number of options for risk-appropriate utilization of CIBCs in various storage and operational scenarios. To improve the management of fire risk, the 2018 edition of NFPA 30 further restricts the use of Composite IBCs in Class II and IIIA liquids to those that are listed and labelled.

Awareness of the fire hazards presented by CIBCs in Combustible Liquid service is low in both the utilization and code enforcement communities. The current framework provided in NFPA 30 (2018), and adopted by a number of state and regional fire codes, provides a reasonably effective risk control framework to manage this hazard to a reasonable level of risk. It is imperative that users and enforcers become better informed of the existing requirements, and conform to them.

New approaches to drainage and containment design provide a new opportunity to consider additional risk control approaches for managing CIBC hazards, rather than heavily relying on the use of fire resistive CIBC container design (as represented by listed and labeled containers). It is anticipated that these new approaches will be more thoroughly debated through the code consensus process in the next revision cycle of NFPA 30.

9 References

[1] United Kingdom Health & Safety Executive. "RR 564 – Fire Performance of Composite IBCs", Available at http://www.hse.gov.uk/research/rrhtm/rr564.htm , Accessed on March 1, 2018.

[2] BBC News, "Chemical Fire Firm Admits Safety Errors", October 2002, http://news.bbc.co.uk/2/hi/uk_news/england/2290631.stm , Accessed on March 23, 2018

[3] Property Insurance Research Group, "The Fire Risk of Intermediate Bulk Containers (IBCs), *National Fire Protection Association*, 2014, http://www.nfpa.org/ibc

[4] Dutch Safety Board, "Investigation Report: Fire at Chemie-Pack Moerdijk", Available at https://www.onderzoeksraad.nl/nl/onderzoek/974/brand-bij-chemie-pack-te-moerdijk-5-januari-2011 , Accessed on March 1, 2018.

[5] Private communication with author

[6] Fire Protection Research Foundation, "Test Reports on CIBC Fire Testing (Scenarios involving empty and full containers)", Available at https://www.nfpa.org/news-and-research/fire-statistics-and-reports/research-reports/hazardous-materials , Accessed on March 1, 2018

[7] United Stated Occupational Safety & Health Administration, "Directive 01-05-014: U.S. Department of Transportation (DOT) Exemption Which Allow Shipment of Combustible and Flammable Liquids in Polyethylene". Available at: https://www.osha.gov/pls/oshaweb/owadisp.show_document?p_table=DIRECTIVES&p_id=1778 , Accessed on March 1, 2018.

[8] National Fire Protection Association, "NFPA 30: Flammable & Combustible Liquids Code". 2018, http://www.nfpa.org/30

[9] FM Global. "Research Prompts New Ignitable Liquid Standard", September 14, 2017, http://www.fmapprovals.com/product-alerts-and-news-events/approved-product-news/approved-product-news-recent-issues/2017/apn-volume-33-issue-2/new-ignitable-liquid-standard

[10] FM Global, "Research Technical Report: Fire Protection Requirements of Empty Intermediate Bulk Containers (IBCs)", July 2012, https://www.fmglobal.com/research-and-resources/research-and-testing/research-technical-reports

Enabling Real-Time Hazard Assessment at the Workplace

Nastassja A. Lewinski
Virginia Commonwealth University
601 W. Main St., Box 843028, Richmond, VA 23284
nalewinski@vcu.edu

Lynn E. Secondo, James K. Ferri
Virginia Commonwealth University
601 W. Main St., Box 843028, Richmond, VA 23284
secondole@vcu.edu, jkferri@vcu.edu

Prepared for Presentation at
American Institute of Chemical Engineers
2018 Spring Meeting and 14th Global Congress on Process Safety
Orlando, Florida
April 22 – 25, 2018

Enabling Real-Time Hazard Assessment at the Workplace

Nastassja A. Lewinski
Department of Chemical and Life Science Engineering
Virginia Commonwealth University
601 W. Main St., Box 843028, Richmond, VA 23284
nalewinski@vcu.edu

Lynn E. Secondo, James K. Ferri
Department of Chemical and Life Science Engineering
Virginia Commonwealth University

Keywords: aerosols, in vitro testing, occupational safety, additive manufacturing

Abstract

Workplace exposures to concentrated aerosols are associated with a number of diseases. When assessing air quality, instruments are available for measuring aerosol concentrations and particle sizes in real-time. By contrast, air contaminants are collected and brought to the laboratory for biological testing. To our knowledge, no instrument is available for testing the toxicity of aerosols in real-time at the site of exposure. This study involves the development of a real-time, portable, exposure cassette for conducting on-site, *in vitro* toxicity testing of workplace air. The compact design also enables targeted personal breathing zone measurements. Data collected on emerging workplace aerosols, such as those generated during 3D printing, are presented.

1 Introduction

Three-dimensional (3D) printing has enabled the creation of metal parts that before were difficult if not impossible to machine. In addition, the rapid prototyping capabilities along with lowered costs [1] have led to the use of 3D printers in a variety of facilities. These include research laboratories in industry and academia as well as a growing number of libraries, schools, and museums in response to the Maker Movement, which focuses on discovery-based learning with hands-on, do-it-yourself projects [2]. Although these spaces started with printers utilizing fused deposition modeling, the rise in 3D printing of hard materials has introduced other modalities including binder-jetting technology, which is an additive manufacturing technology in which a liquid binding agent is selectively deposited to bond powder particles, layer-by-layer, into a solid object. Binder jetting is capable of printing a variety of materials including metals, sands and ceramics. In some cases, printing is followed by heat treatment with concurrent infiltration to produce a high net density part. As the operation of some binder jetting printers involves the handling of loose powders, there is uncertainty regarding the potential for aerosolization and operator exposure.

In this pilot study, different measurement techniques were used to analyze particle counts, sizes, and masses emitted during standard operation of a binder jetting 3D metal printer to verify the safety of the printing process. The emission data is representative of use in a makerspace, such as those located at universities. First time field testing of a newly developed portable *in vitro* exposure cassette, which serves as a human lung cell-based biosensor, was also conducted.

2 Methods

2.1 Printing process

An ExOne R1 three-dimensional printer, utilizing the binder jetting technology developed by Sachs et al. [3], was used as a test platform to test particle exposure during an industrial-grade additive manufacturing process.

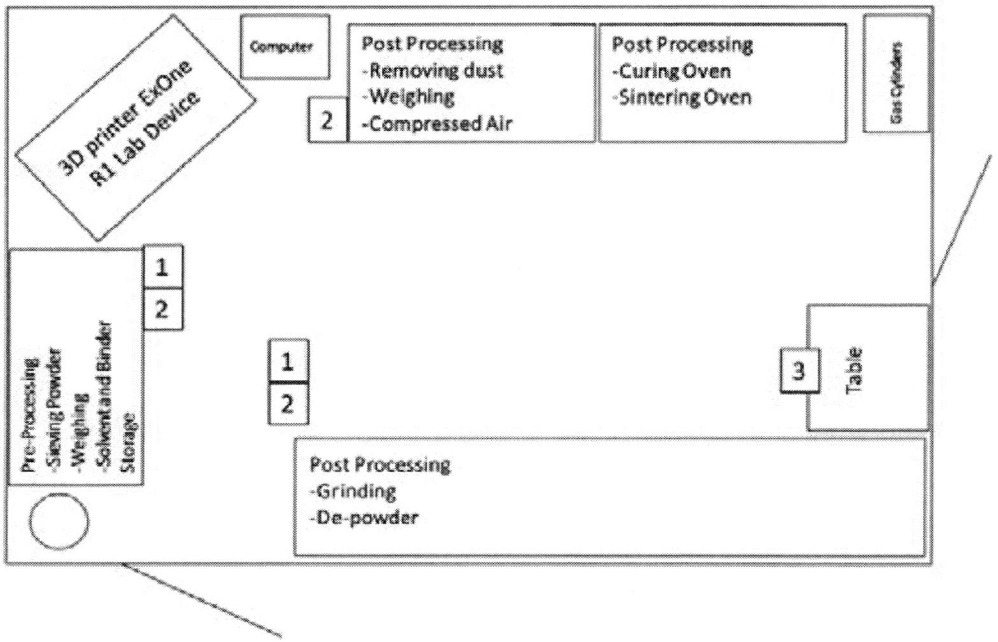

Figure 1. Schematic of printer location and measurement locations (key: 1 = PIVEC, 2 = near-field filter, 3 – far-field filter).

The laboratory housing the ExOne R1 3D printer was located in the Acopian Engineering Center at Lafayette College in Easton, PA. The room was 14’4” by 18’1” with specific areas designated for part pre-processing and post-processing. There were two air return vents and one inlet. Prior to printing, a blank print was performed using the binder solvent and a cardboard coupon. This step checks the functional state of the inkjets to make adjustments before printing. Powder was handled normally as it was added to the build and feed beds, leveled, and brushed to clean the excess powder.

The test prints were performed using the PIVEC design shown in Figure 2A with standard consumable printing materials. For the powder, 30 µm ExOne 420 stainless steel (PM-R1-S4-30) was used. The binder was ExOne BS004 Solvent Binder 04 (PM-B-SR1-04), and the cleaning fluid (PM-C-R1-02) for the print head was CL001 Cleaner. During printing, the powder spread speed was 1 mm s^{-1}, the binder baking time between each printing pass was 30 s and the layer thickness was nominally 100 µm. After print completion, the print envelopment was removed and cured for 2 hours at 200°C intact to enhance the green bonding strength, the adhesive strength of the binder while still wet, of the printed test parts. After curing, the printed part was removed from the unbound powder by manual and pneumatic brushing. In this pilot study, the print process was completed after the curing step. Typically, parts are also sintered using bronze powder to increase the infill within the part. Once sintered any excess powder is cleaned using compressed air and ground to a polished finish.

2.2 *Emissions characterization*

Aerosol generated during the metal 3D printing process was characterized at two locations, within 1 meter (near-field) and ~ 4 meters (far-field) from the source. Mass concentrations were determined by gravimetric analysis. PTFE filters (with support pads, 2 micron pores, SKC Inc.) were stored in standard atmosphere. The filters were static discharged and weighed three times before and after aerosol collection using a BM-22 microbalance (A&D, Japan). Total particulate matter was collected using SKC 37 mm filter cassettes connected to high flow pumps set at a flow rate of 10 liters per minute. PM2.5 was collected using a Parallel Particle Impactor (SKC Inc., USA) connected to a Leland Legacy personal pump with a flow rate of 10 liters per minute. Two filters were collected at each location to measure background and emissions during printing.

Particle number concentrations and particle size distributions were measured using a NanoScan scanning mobility particle sizer (SMPS, model 3910, TSI Inc.) and an optical particle sizer (OPS, model 3330, TSI Inc.). The nanoSMPS measures particles in a size range from 10 to 400 nm nanometers and the OPS measures particles from 0.3 to 10 µm. Transmission electron micrograph (TEM) grids were used to collect particles for TEM imaging.

2.3 *In vitro exposures*

To gain information on the potential effects from inhaling aerosols generated during the metal 3D process, we utilized human lung cells from the bronchial and alveolar regions. The A549, alveolar adenocarcinoma epithelial line, and the BEAS-2B, immortalized bronchial epithelial line, were cultured using complete cell culture media consisting of Dulbecco's Modified Eagle Medium (Gibco) supplemented with 10% fetal bovine serum (Gibco) and 1% penicillin-streptomycin (Gibco). Cells were cultured at the air-liquid interface in cell culture inserts (Corning, PET-track etched, 6.5 mm diameter, 0.4 micron pore), seeded at 35,000 cells/well and 100,000 cells/well, A549 and BEAS-2B cells, respectively, and grown for seven days with media exchange every two days. After seven days, the apical media was removed and the cells were allowed to equilibrate for at least 24 hours prior to exposure. The cells were prepared for transport to Lafayette College by aspirating the media and adding 1% agarose media to each well. The plate was kept at room temperature until the agarose solidified around the cell culture insert. After transport, the cells were returned to normal culture and equilibrated for 24 hours.

The Portable In Vitro Exposure Cassette (PIVEC) [4] was used to expose the lung cells by pulling air through the device using personal air sampling pump protected with a HEPA filter. The PIVEC (Figure 2) was designed to house a cell culture insert with culture media and facilitate direct exposure of cells to aerosols.

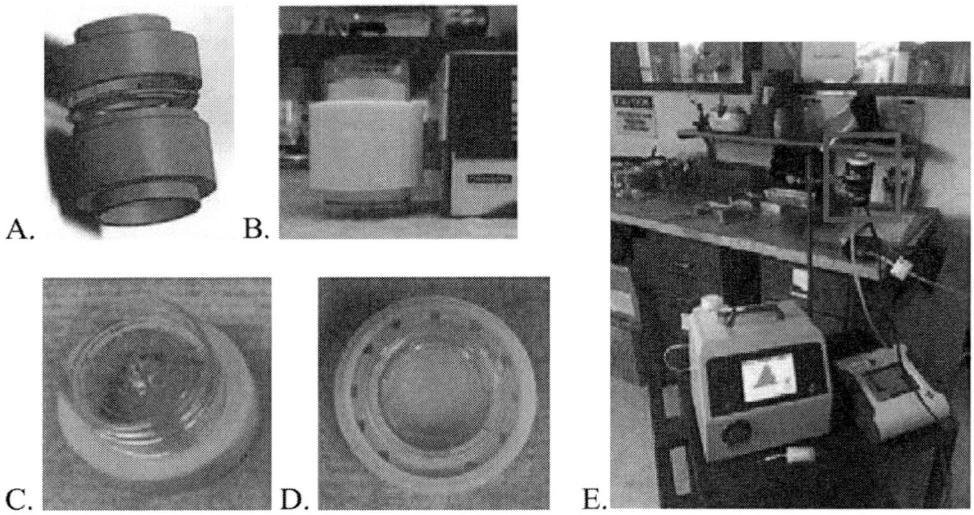

Figure 2. PIVEC – A. AutoCAD drawing of PIVEC B. side view of assembled PIVEC, C. top view of PIVEC aerosol entrance, D. top view of PIVEC cell culture insert housing E. experimental set up, PIVEC (boxed in red) with nanoSMPS (left) and OPS (right) particle counters near metal print post-processing station.

Prior to exposure, cells were pre-loaded with the dichloro-dihydro-fluorescein diacetate (DCFH-DA) indicator dye by replacing the basolateral culture media with 0.5 mL of 10 μM DCFH-DA in Hank's Balanced Salt Solution. The plate was then covered with aluminum foil to prevent photo-activation and incubated at 37°C for 60 minutes. After removing the DCFH-DA solution, the cell culture inserts were transferred to a new 24 well plate containing 0.5 mL of HBSS buffer in each well, and fluorescence was measured on a microplate reader (Biotek Synergy) at $\lambda ex/\lambda em$ = 485/530 nm to determine the pre-exposure fluorescence intensity. The cells were then loaded into the PIVEC and the PIVEC was connected to a pump (Airchek Touch, SKC Inc.) set at a flow rate of 0.5 liters per minute. The cells were exposed in the near-field for one hour. After exposure, intracellular reactive oxygen species (ROS) generation using the DCFH-DA assay and cell viability using the lactate dehydrogenase (LDH) assay were assessed. The cell inserts were transferred to a 24 well plate and read at $\lambda ex/\lambda em$ = 485/530 nm to determine the ROS production immediately after and 1 hour post-exposure. From the basolateral buffer solution, triplicate samples consisting of 100 μL were transferred to a white 96-well plate, and to each sample well, 100 μL of CytoTox-ONE Reagent (Promega) was added. The plate was incubated at room temperature for 10 minutes after which 50 μL of stop solution was added to each well. The plate was shaken and read on a microplate reader at $\lambda ex/\lambda em$ =570/620 nm to determine the cell viability immediately after exposure.

3 Results

3.1 Aerosol concentration and size distribution

Figure 3 shows the particle number concentrations measured using an optical particle counter for the duration of the experiment. The average background concentration was 3.3 x 10^4 particles/cm^3. The average particle concentration changed to 3.5 x 10^4 particles/cm^3 during printer setup, 3.8 x 10^4 particles/cm^3 during printing, and 3.3 x 10^4 particles/cm^3 during part de-powdering. Despite measuring a slight elevation in concentration, these changes are not statistically significant as compared to background.

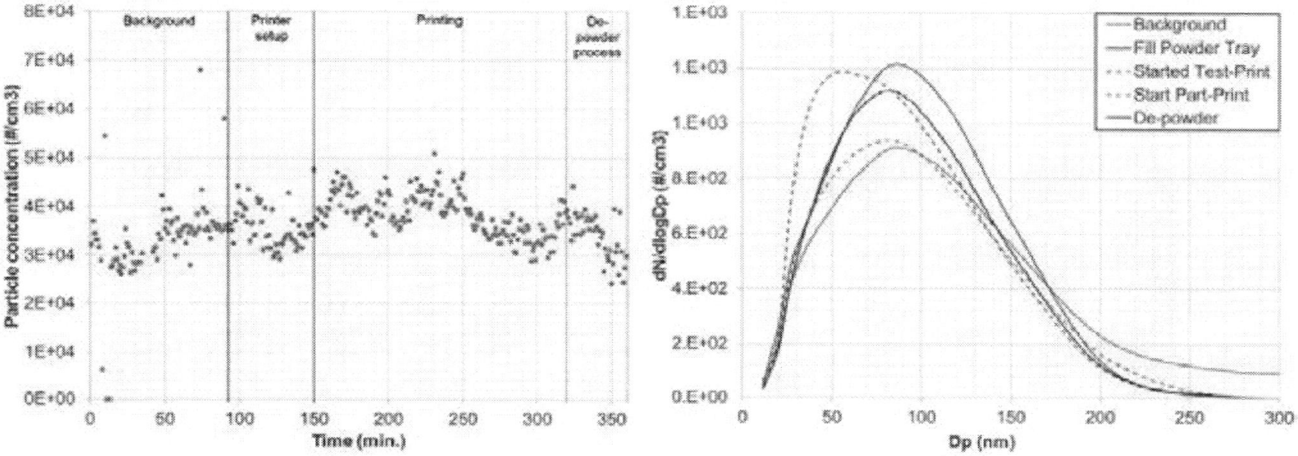

Figure 3. Total particle number concentration (left) and average particle number distributions (right) during the 3D printing process.

Particle size as measured using a scanning mobility particle sizer revealed a geometric mean diameter of 87 nm for the background, and during powder tray filling, test print and part de-powdering. Interestingly, a geometric mean diameter of 65 nm was measured during the part print (Figure 3, right).

3.2 Gravimetric mass

Total particulate matter was collected on PTFE filters to determine both background and process-related aerosol mass concentrations. Two hour collection durations did not provide significant amounts of gravimetric deposition. Six hour collection durations yielded total aerosol measurements of 0.08 mg/m^3 ($PM_{2.5}$ 0.03 mg/m^3) within 1 meter of the printer and 0.02 mg/m^3 more than 3 meters from the printer with no difference between background and during print conditions.

3.2 *In vitro biological response*

Two controls were measured, one where the cells were kept in normal culture conditions, i.e. incubated at 37°C, with 5% CO_2 and a relative humidity > 80%, and one where the cells were exposed to HEPA-filtered air at 0.5 liters per minute using the PIVEC. As shown in Figure 4, some reduction in cell viability was measured after exposure to HEPA-filtered air due to the continuous airflow and lower relative humidity (50%). Further reduction in cell viability was measured in the samples exposed during parts of the metal 3D printing process, i.e. test print and de-powder (Figure 4, left). No statistically significant changes in intracellular ROS generation were measured in the A549 or BEAS-2B cells immediately after exposure as compared to the negative controls. Although an increase in ROS generation was measured 1 hour post-exposure in the incubator control and HEPA filtered air control, the increase in fluorescence intensity is attributed to normal cellular respiration.

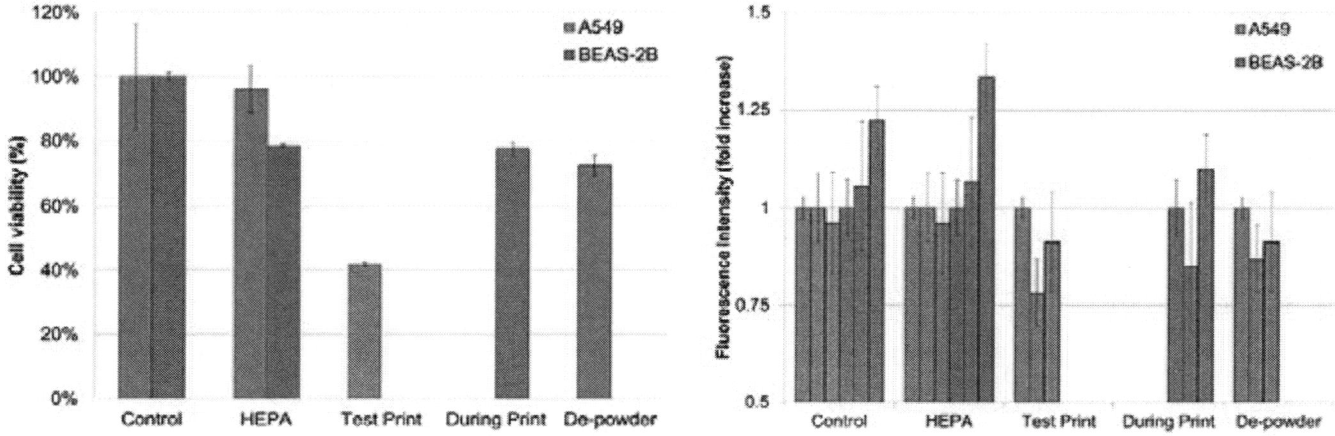

Figure 4. Cell viability (left) and intracellular ROS generation (right, key: left bar = before exposure, middle bar = immediately after exposure, right bar = 1 hour post-exposure) measurements, N = 2.

4 Discussion

In this work, we presented preliminary emission and toxicity assessment results collected from a metal binder jet printing process. When following the standard operating procedure recommended by the manufacturer of the metal 3D printer, small fluctuations in aerosol particle number concentration were measured. Little to no reduction in cell viability was measured in lung cells exposed to aerosols during the print and de-powder steps as compared to cells exposed to HEPA filtered air. Although reduction in viability was observed in cells exposed during the pre-print step, replicate experiments are necessary to verify this preliminary finding.

Limited data is available in the literature on particulate matter emissions from binder jetting processes. To our knowledge, no emissions data is available in the literature for metal binder jet printing. Afshar-Mohajer et al. examined emissions from binder jet printing with a polymer composite powder [5]. During the printing process, particles between 205-255 nm in size were reported in greatest particle number concentration (0.9-1.16 x 10^4 particles/cm^3). The highest aerosol concentrations (~4.4 x 10^4 particles/cm^3) were measured when the printer cover was removed to eject the print.

In addition to binder jetting, other metal-based additive manufacturing techniques are available, including selective laser sintering or melting. Graff et al. reported preliminary emission measurements from selective laser melting of a metal powder containing nickel, chromium, and cobalt [6]. The particle sizes of the metal powder ranged from 15-45 μm. The particle size of the powder used in this study was also within this size range (30 μm). Both are larger than the 10 μm limit recognized for the inhalable fraction of aerosols [7]. Although peaks in particle number concentration were reported by Graff et al., the concentrations in their study and this study are small in comparison with other nanoparticle aerosol generating processes, such as tungsten inert gas welding [8]. Because the metal powder used by Graff et al. contained known toxic metals, the authors recommended biomonitoring in addition to air monitoring [6]. This recommendation is in line with our proposed approach to assess the biological responses to contaminants in workplace air using a cell-based biosensor.

Few published studies report the assessment of *in vitro* biological responses after direct exposure to outdoor aerosols [9-10]. There is rising interest in improving workplace air quality since lower contaminant concentrations have been associated with increased cognitive function [11]. Therefore more research examining the *in vitro* biological responses after direct exposure to workplace indoor aerosols is needed. Our study examined only one example of an additive manufacturing process that could generate aerosolized nanoparticles at the workplace. Numerous reports indicate nanoparticle aerosol generation by fused deposition modeling based 3D printing [12]. With inclusion of active pharmaceutical ingredients in the case of 3D printed drugs, a small increase in aerosol concentration during the process could be significant [13]. In addition to additive manufacturing, established processes, such as welding, grinding, and bagging operations, could also benefit from biological response monitoring using technologies such as the PIVEC presented here.

5 Conclusion

When operated as intended, little to no elevation in aerosol concentrations were measured during different steps in the metal 3D printing process using a binder jetting system. Preliminary *in vitro* measurements suggest that small changes in cell viability and oxidative stress resulted from direct exposure to room air sampled near the printer and near the de-powder workbench. Future work is needed to examine aerosol generation when multiple printers are in operation as well as use of non-standard metal powders, e.g. nanoscale powders. Our portable *in vitro* exposure cassette offers a new approach to monitoring workplace air quality.

6 References

[1] D.S. Thomas and S.W. Gilbert. "Costs and cost effectiveness of additive manufacturing," NIST Special Publication, 1176, p.1-89, 2014.

[2] D. Dougherty. "The Maker Movement," innovations, 7 (3), pp. 11-14, 2012.

[3] E. Sachs, M. Cima, P. Williams, D. Brancazio, and J. Cornie. "Three dimensional printing: rapid tooling and prototypes directly from a CAD model," Journal of Engineering for Industry, 114 (4), pp. 481-488, 1992.

[4] L.E. Secondo and N.A. Lewinski. "A new portable in vitro exposure cassette for aerosol sampling," submitted, 2018.

[5] N. Afshar-Mojajer, C.Y. Wu, T. Ladun, D.A. Rajon, and Y. Huang. "Characterization of particulate matters and total VOC emissions from a binder jetting 3D printer," Building and Environment, 93, pp. 293-301, 2015.

[6] P. Graff, B. Stahlbom, E. Nordenberg, A. Graichen, P. Johansson, and H. Karlsson. "Evaluating Measuring Techniques for Occupational Exposure during Additive Manufacturing of Metals: A Pilot Study," Journal of Industrial Ecology, 21 (S1), pp. S120-S129, 2017.

[7] J.S. Brown, T. Gordon, O. Price, and B. Asgharian. "Thoracic and respirable particle definitions for human health risk assessment," Particle and Fibre Toxicology, 10 (1), pp.12, 2013.

[8] H. Graczyk, N. Lewinski, J. Zhao, N. Concha-Lozano, and M. Riediker. "Characterization of Tungsten Inert Gas (TIG) Welding Fume Generated by Apprentice Welders," Annals of Occupational Hygiene, 60 (2), pp. 205-219, 2016.

[9] W. Vizuete, K.G. Sexton, H. Nguyen, L. Smeester, K.M. Aagaard, C. Shope, B. Lefer, J.H. Flynn, S. Alvarez, M.H. Erickson, and R.C. Fry. "From the Field to the Laboratory: Air Pollutant-Induced Genomic Effects in Lung Cells," Environmental Health Insights, 9, pp. EHI-S15656, 2015.

[10] C. Bisig, A. Petri-Fink, and B. Rothen-Rutishauser. "A realistic in vitro exposure revealed seasonal differences in (pro-)inflammatory effects from ambient air in Fribourg, Switzerland," Inhalation Toxicology, 30 (1), pp. 40-48, 2018.

[11] J.G. Allen, P. MacNaughton, U. Satish, S. Santanam, J. Vallarino, and J.D. Spengler. "Associations of cognitive function scores with carbon dioxide, ventilation, and volatile organic compound exposures in office workers: a controlled exposure study of green and conventional office environments," Environmental Health Perspectives, 124 (6), pp.805-812, 2016.

[12] P. Azimi, D. Zhao, C. Pouzet, N.E. Crain, and B. Stephens. "Emissions of ultrafine particles and volatile organic compounds from commercially available desktop three-dimensional printers with multiple filaments," Environmental Science & Technology, 50 (3), pp.1260-1268, 2016.

[13] J. Norman, R.D. Madurawe, C.M.V. Moore, M.A. Khan, and A. Khairuzzaman. "A new chapter in pharmaceutical manufacturing: 3D-printed drug products," Advanced Drug Delivery Reviews, 108 (1), pp. 39-50, 2017.

Pollution, Control Status and Strategy of Volatile Organic Compounds in China's Petrochemical and Refining Industries

Jia Ke, China University of Petroleum (East China), Qingdao, P R China
Dongfeng Zhao, China University of Petroleum (East China), Qingdao, P R China; Center for Safety, Environmental, and Energy Conservation Technology, Qingdao, P R China
Yi Liu, China University of Petroleum (East China), Qingdao, P R China; Center for Safety, Environmental, and Energy Conservation Technology, Qingdao, P R China

Abstract

With rapid industrialization and urbanization in China, haze and photochemical smog, caused by fine particulate matters ($PM_{2.5}$, PM_{10}) and ozone (O_3), have become one of the most severe environmental pollution issues in China, which were seriously affecting the daily life and physical health of residents. Thus, as an important precursor of fine particulate matters and ozone, volatile organic compounds (VOCs) are now included in the national control of key pollutants and total control factors. Because of its strategic importance in industrial system and its huge VOCs emissions (accounts for 34.3% of the total emissions), petrochemical and refining industries have recently elicited public concern and given new impetus to scientific interest.

This paper reviews the VOCs emissions, and summarizes the policies, regulations, emission standards and the existing pollutant permit system for controlling VOCs in China's petrochemical and refining industries. This paper also analysis the main problems in current petrochemical and refining industries VOCs control status and management. Some countermeasures and suggestions, such as developing and improving VOCs monitoring methods, establishing VOCs emission list database, researching the best feasible control technology in petrochemical and refining industries and so on, are proposed to strengthen the control and management of VOCs emissions. Besides, the prospects and recommendations are put forward for China's future open, whole-process VOCs management model. The findings not only can provide a scientific basis for enhancing and implementation of existing VOC control strategies, but also can provide technical support for governments and enterprises to improve environmental quality and alleviate regional air pollution.

Keywords: Petrochemical and Refining Industries; Volatile Organic Compounds; Policies; Environmental Management

Introduction

With the rapid development of China's industrialization process, air pollution type has shifted from traditional coal smoke pollution, which dominated by dust and SO_2 emissions, to a composited air pollution, which dominated by fine particulate matters ($PM_{2.5}$, PM_{10}) and ozone (O_3) pollution. Air pollution is increasingly intensifying. The continuous phenomena of acid rain, haze and photochemical smog have become one of the most serious environmental issues in many cities and regions in China, seriously affecting people' daily life and physical health.

Many articles show that volatile organic compounds (VOCs) are important precursors of atmospheric ozone (O_3) and secondary organic aerosol (SOA). SOA is considered to be one of main contribution sources to $PM_{2.5}$ in the ambient air, which has contribution to the formation of the composited air pollution. Among various VOCs anthropogenic emission sources, industrial sources are the largest source of VOCs emissions. The petrochemical and refining industries, which has significant VOCs emissions among industrial sources, is undoubtedly placed on the front line of VOCs control. Therefore, VOCs research and effective control in the petrochemical and refining industries has become a hot topic in the field of atmospheric environmental management and control in China.

The status of VOC emissions in the petrochemical and refining industries, including policies, regulations, standards of VOCs control and management, were analyzed and studied. Some main problems in VOCs control current status and control management were proposed. In addition, solutions and recommendations were provided. The results can will provide technical supports to improve environmental quality.

1 China Air Quality Status

Nowadays, China has established a comprehensive national environmental air quality monitoring network and has set up more than 5,000 monitoring stations, which included 1,436 air quality monitoring stations in 338 cities at or above prefecture level (contained 135 clean control stations), 96 regional (rural) ambient air quality monitoring stations, 16 background ambient air quality monitoring stations, and 38 atmospheric particulate components/photochemical monitoring stations in 2+26 cities (i.e. Beijing, Tianjin, and 26 cities in the Beijing-Tianjin-Hebei and surrounding area).

According to the Ambient Air Quality Standards (GB 3095-2012) and the Technical Regulation for Ambient Air Quality Assessment (HJ 663-2013), six ambient air quality pollutants, i.e. SO_2, NO_2, $PM_{2.5}$, PM_{10}, O_3 and CO, were evaluated based on the ambient air quality monitoring data. The ratio of different air quality levels was assessed in different cities from 2013 to 2016. And its results were showed in Table 1. The average annual concentrations of ambient air quality pollutants from 2013 to 2016 were shown in Table 2.

Table 1 Air quality conditions from 2013 to 2016

Year	percentage of air quality standards				
	excellent and good	slightly polluted	intermediately polluted	heavily polluted	very heavily polluted
2013	60.5	22.9	8	6.2	2.4
2014	66.3	21.6	6.5	4.4	1.2
2015	76.7	15.9	4.2	2.5	0.7
2016	78.8	14.8	3.7	2	0.6

Note: Days with AQI = 0~100 indicates perfect, AQI =101~150 indicates slight pollution, AQI =151~200 indicates intermediate pollution, AQI =201~300 indicates heavy pollution and AQI >300 very serious pollution.

Table 2 Average annual concentrations of ambient air quality pollutants from 2013 to 2016

Year	$PM_{2.5}$	PM_{10}	90th percentile concentration of O_3 daily maximum 8-hour average	SO_2	NO_2	95th percentile concentration of daily CO average (mg/m^3)
2013	72	118	139	40	44	2.5
2014	62	105	140	35	38	2.2

2015	50	87	134	25	30	2.1
2016	47	82	138	22	30	1.9
Standard valve	35	70	160	60	40	4

The percentage of cities met national air quality standard was increased from 4.1% in 2013 to 24.9% in 2016, and the percentage of cities failed to meet national air quality standard was decreased from 95.9% in 2013 to 75.1% in 2016. By comparing the ambient air qualities data from 2013 to 2016, it can be found that the urban air quality has improved as a whole. It indicated that China's air pollution prevention and control has made significant progress under control and management. The annual average concentrations of $PM_{2.5}$, PM_{10}, NO_2, SO_2, and CO decreased year by year. The number of heavy pollution days in most cities has decreased. However, pollution issues, such as high-pollution loads of $PM_{2.5}$ and PM_{10} in key areas, composited pollution and heavy air pollution days which the primary pollutants were $PM_{2.5}$, O_3 and PM_{10} during these days, are still serious. The overall situation of air pollution control is still severe.

As an important precursor of SOA and ozone, VOCs have become the fourth most important atmospheric pollutants in China after sulfur dioxide, nitrogen oxides, and dust, and have been listed in national key prevention and control pollutants and total control factors. Therefore, the control and reduction of VOCs has also become one of the essential issues in the prevention and control of atmospheric pollution. In addition, petrochemical industry has attracted much attention due to its huge VOCs emission, and has become one of the key industries for VOCs management and control.

2. Volatile Organic Compounds

Volatile Organic Compounds (VOCs) are composed of hundreds of organic compounds that are directly emitted into the atmosphere. They are a class of organic compounds. For different industries, the chemical composition varies, but mainly includes organic compounds such as alkanes, unsaturated hydrocarbons, halogenated hydrocarbons, benzene series, organic acids, alcohols, and esters. For volatile organic compounds, the United States, the European Union, Japan, and other countries have been research and control VOCs decades ago. However, China starts relatively late to research and control VOCs, and is still at exploratory stage.

2.1 Definition of VOCs

There is no formal and standard definition of VOCs in the world nowadays. In china, many standards also defined VOCs, but it existed difference. The specific definitions were described in Table 3 below.

Table 3 Definition of VOCs in different standards

No.	Standard	Definition
1	Emission standard of pollutants for synthetic leather and artificial leather industry GB 21902-2008	Any organic compound (excluding methane) that has a boiling point of less than 250°C or can be discharged into the air as a gaseous molecule at atmospheric pressure
2	Emission standards of air pollutants for petroleum refining and petrochemicals manufacturing industry DB 11/447-2007	Organic compounds with a vapor pressure higher than or equal to 0.01 kPa at 20°C
3	The discharge standard of pollutants	Volatile organic compounds that saturated vapor

	for bio-pharmaceutical industry DB 31/373-2010	pressure at 0.1mmHg or above at 20°C, or melting point below room temperature and boiling point below 260°C, excluding methane
4	Indoor air quality standard GB/T 18883-2002	Volatile organic compounds that retention time between hexane and n-hexadecane analyzed by non-polar columns (polarity less than 10), which sampled by Tenax GC or Tenax TA
5	Integrated emission standard of air pollutants GB 3095-2012	Organic compounds with a vapor pressure higher than or equal to 0.01 kPa at 20°C or has corresponding volatility under specified conditions
6	Emission standard of pollutants for petroleum refining industry GB 31570-2015	Organic compounds involved in atmospheric photochemical reactions, or determined or calculated by specific methods
7	Emission standard of pollutants for petroleum chemistry industry GB 31571-2015	Organic compounds involved in atmospheric photochemical reactions, or determined or calculated by specific methods

As can be seen from Table 1, there are different definitions of VOCs in different standards. These definitions were defined in terms of volatility (such as vapor pressure, boiling point, etc.). Under this definition, VOCs basically cover all organic compounds. However, it is not clear that which substances have O_3 and $PM_{2.5}$ formation potential. They have limitations. In addition, there are some definitions that are defined based on monitoring methods. Under this definition, some of the higher melting VOCs are excluded from the definition, which cannot fully reflect the definition.

For petrochemical and refining industries, the definition is mainly based on photochemical reactivity. Under this definition, the factors can be found to control VOCs, so as to achieve the control of the formation of ozone in ambient air. It is also possible to find out specific substances with high photochemical activity at the source stage. The formation of ozone and $PM_{2.5}$ in the ambient air can be achieved by not using or instead of using such substances, or developing new processing technologies.

2.2 Sources of VOCs

The sources of VOCs include natural sources and human sources. Natural sources mainly come from plant emissions, such as plant releases, volcanic eruptions, and forest fires. They are uncontrolled. The most important emissions sources are forests and shrubs. The most important natural emissions are isoprene and monoterpenes. The human sources mainly come from the use of automobile exhaust, fuel evaporation, petrochemicals, and paints. It includes various industrial sources, agricultural sources, transportation sources, and life sources.

In China, human sources mainly come from industrial sources, transportation sources (motor vehicle emissions) and biomass combustion. Due to the rapid industrialization process, VOCs emissions from industrial sources increased year by year and accounted for more than 50% of the total VOCs emissions. Industrial emission sources are complex, involving production, usage, storage, and transportation processes. The petrochemical and refining, coatings, inks, pesticides, and automotive industries accounted for more than 80% of total industrial VOCs emissions. Research showed the VOCs emissions from petrochemical and refining industry represented about 34.3% of the total VOCs emissions.

2.3 Photochemical Reaction of VOCs

Under UV light, VOCs have a series of photochemical interactions with nitrogen oxides (NOx), which formed O_3, peroxyacetyl nitrate (PAN), highly reactive free radicals (ON, RO_2, HO_2), aldehydes and so on. These pollutants can further form secondary aerosols (SOA) through oxidation, nucleation, and condensation processes, which promotes the formation of $PM_{2.5}$.

When there is no VOCs in the air, the interactions between NO_2, NO, and O_3 in the atmosphere can reach equilibrium:

$$NO_2 \rightarrow NO + O$$

$$O + O_2 \rightarrow O_3$$

$$O_3 + NO \rightarrow NO_2 + O_2$$

When there is no VOCs in the air, the following reactions occur:

$$VOCs + OH \rightarrow HO_2$$

$$HO_2 + NO \rightarrow NO_2 + OH \text{ (reaction quickly)}$$

NO, produced by NO_2, is consumed by HO_2, resulting in NO being unable to react with ozone and breaking the equilibrium of NO_2, NO, and O_3 interactions. It eventually leading to the accumulation of O_3.

2.4 Damages of VOCs

Due to the diversity of VOCs' physical and chemical properties, the scope and impact of VOCs are also relatively widespread. VOCs are not only harmful to environmental, but also harmful to human health.

1) Impact on the environment

Some of the substances of VOCs have strong photochemical reactivity, can cause photochemical smog, and can indirectly affect the radiation balance on the Earth's surface, which in turn affects the climate. Some VOCs can destroy stratospheric ozone characteristics and reduce stratospheric ozone concentrations, such as chlorofluorocarbons (CFCs). With the diffusion and migration of VOCs pollution, the environmental problems such as acid rain, ozone layer destruction and atmospheric warming are no longer belonged to one city or country, but spread to a global scale.

2) Impact on human health

Most VOCs are toxicologically and harmful to human health. At present, more than 50 pollutants are contained in priority control pollutants list, which announced by the US EPA. VOCs can cause maladjustment of multiple systems, such as the respiratory system, nervous system and cardiovascular system. VOCs also have carcinogenicity and genetic toxicity. When the concentration of VOCs in the environment reaches a certain concentration, people may feel headaches, nausea, and vomiting. In severe cases, convulsions, coma, and even life-threatening conditions may occur.

3. Status of VOCs Emissions in Petrochemical and Refining Industries

As a pillar industry for the economy development of China, petrochemical and refining industries arises a series of environmental pollution problems, especially VOCs pollution problems during the acceleration of industry economic growth. Researchers used different estimation methods to estimate the VOCs emissions. The VOCs emissions from human-made source in China are about 20 Mt/a, and the data rapidly rising. Among the human-made source, industrial sources accounted for about 55%, in which petroleum refining and storage, organic chemicals, and fine chemicals were key industries for VOC emissions.

In November 2015, the Ministry of Ecology Environment (MEE) issued the "Guidelines for

VOCs Pollution Source Inspection Work for petrochemical and refining industries" in accordance with the theoretical basis for the classification and analysis of pollution sources in the process chain, and divided VOCs sources into 12 categories. These are the leakage of equipment static and dynamic sealing points; the volatilization loss of organic liquids storage and blending process; the volatilization loss of organic liquids loading and unloading process; the volatilization loss of wastewater in collection, storage, treatment and disposal process; organized emissions; the release from circulating cooling water system; abnormal conditions (including startup, shutdown and maintenance) emissions; flares emissions; combustion emissions; unorganized emissions; sampling process emissions and accident emissions. The document requires that all petrochemical and refining industries should basically complete the VOCs comprehensive control and establish VOCs monitoring system by 2017. The total VOCs emissions will be reduced by more than 30% from 2014. By the end of 2017, the petrochemical and refining industries have comprehensively carried out comprehensive VOCs control work.

The VOC emissions of each source from petrochemical and refining industries with different capacities are shown in Table 4.

Table 4 VOC emissions from different petrochemical and refining industries (t/a)

Source	**Sinopec Qingdao Refining Corp., Ltd.**	**Haerbin Petrochemical**	**Guangxi Petrochemical**	**Jiujiang Petrochemical**
Capacity (million t/a)	1000	420	1000	800
Leakage of equipment static and dynamic sealing points	90.2	212.9	126.3	199.4
Volatilization loss of organic liquids storage and blending process	319.1	997.3	2087.8	2027.8
Volatilization loss of organic liquids loading and unloading process	26.1	511.9	1613.9	1084.4
Volatilization loss of wastewater in collection, storage, treatment and disposal process	228.4	130.5	159.4	1061.5
Release from circulating cooling water system	225.6	124.3	121.4	853.2
Combustion emissions	7.5	19.4	221.3	13.8
Organized emissions	48.6	2.6	370.2	87.97
Unorganized emissions	472.5	0.0	0.0	167.5
Sampling process emissions	0.0	0.0	19.9	0
Flares emissions	7.2	143.6	81.7	0.02
Abnormal conditions emissions	0.0	0.0	3.0	0.4
Accident emissions	0.0	0.0	0.0	0
Total	1425.1	2142.4	4805.0	5495.9

4 VOCs Control Status

Facing the rising industrialization and deteriorating ambient air quality, China has begun to pay attention to VOCs pollution, and has gradually established a management idea, taking whole process and refinement management and control as center. VOCs management and control measures will be conducted from the source, process, and ends aspects. In an attempt to reduce air pollution and control VOCs pollution, the Chinese government has made the decision to enforce stricter regulations, including laws, standards, regulation, action plans, VOCs pollution prevention technologies and others.

4.1 Policy System

For air pollution, the two key laws are Environmental Protection Law (EPL) and Law of the People's Republic of China on the Prevention and Control of Atmospheric Pollution. As the air pollution has been shifting from typical coal smoke pollution to composited air pollution, the two laws are amended on April 2104, August 2015 respectively and implemented the next year. The current version of Prevention and Control of Atmospheric Pollution Law has expanded into the scope of legal supervision, and defined legal responsibilities of VOCs prevention and control.

Since 2010, in order to govern VOCs and improve environmental air quality, the State has successively issued various regulations and action plans, such as 'Guiding Opinions on Promoting Air Pollution Joint Prevention and Control Work to Improve Regional Air Quality', '12th Five-year Plan' on Air Pollution Prevention and Control in Key Regions', 'the State Council Air Pollution Prevention and Control Action Plan', 'Control Pollutant Release Permit System Implementation Plan'. These regulations and action plans listed VOCs as key pollutants for key air pollution prevention and control, together with particulate matter, SO_2, and NO_2. They are also placed $PM_{2.5}$ and O_3 and its precursors in a core position in the prevention and control of air pollution in China. It is proposed that by 2020, the state will establish and improve the VOCs pollution prevention and management system, and implement VOCs pollution reduction in key areas and key industries.

As one key industry, petrochemical and refining industries, the MEE issued the 'Comprehensive Plan for the Control of Volatile Organic Compounds in the Petrochemical and Refining Industries', the 'Guidelines for VOCs Pollution Source Investigation in the Petrochemical and Refining Industries' and the 'Guidelines for Petrochemical and Refining Industries Leak Detection and Restoration Work' and other documents to comprehensively control VOCs in the petrochemical and refining industries, and to greatly reduce VOCs emissions from the petrochemical and refining industries.

4.2 Emission Standards

To strengthen the control of atmospheric pollution, the National Ambient Air Quality Standard (NAAQS) (GB 3095-2012), which newly limits the ambient levels of $PM_{2.5}$ and ozone and sets clear reduction goals for air quality planners and regulators, was revised on 29 February 2012 and enacted on 1 January 2016. As for petrochemical and refining industries, some industrial emission standards have issued, like 'Emission Standard of Pollutants for Petroleum Refining Industry' (GB 31570-2015), 'Emission Standard of Pollutants for Petroleum Chemistry Industry' (GB 31571-2015) and so on.

In addition to the national standards, various local standards have been established. For example, Beijing has established the 'Integrated Emission Standard for Air Pollutants' (DB 11/501-2007) and the 'Emission Standard of Air Pollutants for Refining and Petrochemical Industry' (DB 11/447-2007). Shanghai has newly established the 'Integrated Air Pollution Emission Standard' (DB 31/933-2015). The standards and documents established by various provinces have further strengthened the control of VOCs pollution.

Furthermore, the state has also put forward more stringent emission standards for vehicles. Since 2017, the national vehicle emission pollutants will fully implement the national V emission standards. For petrochemical and refining industries, the more stringent gasoline and diesel standards enables petrochemical enterprise to further optimize the oil processing technology and reduce the sulfur and benzene concentration in petroleum products. To a certain extent, it also can reduce the VOCs emissions from petrochemical and refining industries.

4.3 Regional Joint Prevention and Control

The complex air pollution in China is not just a local issue but a domestic regional cross-boundary problem. In order to solve this problem, regional coordination to control cross-boundary air pollutants is necessary to reduce the impact of atmospheric pollution on downwind locations, and improve the regional air quality.

Until now, nine regions, including the Beijing-Tianjin-Hebei region, the PRD region, the YRD region, central Liaoning region, Shandong Peninsula, Western Taiwan Straits Metropolitan region, Cheng-Yu Metropolitan region, Wuhan Metropolitan region, and Cheng-Zhu-Tan region, have been designated as pilot regions for the joint prevention and control of regional air pollution.

The effectiveness of regional joint prevention and control is gradually emerging. For example, on 20 August 2015, ahead of the 70th-anniversary celebrations of the end of World War II, the Beijing and surrounding areas shut down industrial facilities and reduced car emissions. This action resulted in $PM_{2.5}$ concentration lower than the national air quality standard. The restrictions resulted in an average Beijing $PM_{2.5}$ concentration of 19.5 $\mu g/m^3$, the lowest that had ever been on record in the capital, together with the Parade Blue phenomena. At the same time, various local governments (such as Beijing, Shanghai, Guangdong, etc.) are also developing different policies and plans to conduct VOCs prevention and control.

Besides, in order to improve the air quality in autumn and winter, the state has new requirements for new and existing enterprises in "2+26" cities (including Beijing, Tianjin, and 26 cities in the Beijing-Tianjin-Hebei and surrounding area). The pollutants in these enterprises will implement specific emission limits from March 1, 2018 and October 1, 2018 respectively.

4.4 VOCs Prevention Technologies

VOCs pollution prevention and control follows the principle of comprehensive prevention and control combined with source control, process control and end treatment.

Source control mainly refers to develop and use less contaminated raw materials, eliminate materials which have many types of VOCs and volatility, and use alternatives, alternative technologies. For petrochemical and refining industries, adopting advanced clean technologies, improving the conversion and utilization efficiency of crude oil are encouraged. Reducing or avoiding the emission of VOCs at the source is the primary way to control VOCs pollution. It is easy for VOCs to violate and leak during usage, transportation, and storage. Process control can be achieved by controlling leakage, collecting VOCs and so on. In addition, it also can be achieved by taking measures, such as improving the automation and sealing of equipment components, optimizing process, improving equipment, and using clean process, to reduce VOCs emissions.

End treatment is the final step to control VOCs emissions, which is the research focus and hotspots. Based on the technology principles, the end treatment techniques of VOCs can be broadly divided into two categories: decomposition and destruction technologies and recycling technologies. On the basis of investigation of VOCs pollution control technologies status, the MEE issued the 'National Advanced Pollution Prevention and Control Technology Catalogue (VOCs Prevention and Control Technology)', listing various proposed advanced VOCs end treatment technologies and its application scope (below table 5 only lists processing technologies involving the petrochemical industry). These technologies are innovative. The effects are stable and the cost is reasonable. The technologies should be encouraged for enterprises to promote applications

Table 5 Advanced VOCs end treatment technologies and its application scope

No.	Technology	Application scope
1	Activated carbon adsorption-nitrogen desorption condensation solvent recovery technology	Petrochemical and other industries
2	Oil and Gas Active Carbon Adsorption and Recovery Technology during Oil Storage and Transportation	Oil and gas recovery during product oil loading or storage
3	Oil and Gas Membrane Separation and Adsorption Recovery Technology for Oil Storage and Transportation	Petrochemical industry oil and gas recovery
4	Stationary organic waste heat storage combustion technology	Medium or high-concentration VOCs exhaust gas purification in petrochemical and other industries
5	Rotary heat storage combustion purification technology	Medium or high-concentration VOCs exhaust gas purification in petrochemical and other industries
6	Regenerative catalytic combustion (RCO) technology	Medium and high concentration VOCs exhaust gas treatment
7	Low concentration organic waste gas biological purification technology	Low concentration organic waste gas treatment
8	Advanced oxidation-biological purification coupled processing technology	Purification of low-concentration VOCs and malodorous gases from the production processes and sewage treatment plants (stations) in industries such as petroleum refining

4.5 Pollutant Permit System

China's pollutants permit system began in the 1980s, but it did not play a central role at that time. Until November 2016, the new pollutant permit system began to be implemented nationwide. It was clearly proposed that by 2020, the pollutants permit covering all stationary pollution sources should be issued, and a permit emission control system should be established, and impelling the "one-signature" management. In recent years, the state has issued 'Provisional Regulations on the Management of Emission Permits', 'Administrative Measures on Emission Permits (Trial)' and other documents to present detail provisions on the procedures for the application, review, issuance, and management of pollutant permit. At the same time, regulations on the allowable emission concentrations, permitted emissions, and environmental management for air pollutants were also provided.

For the petrochemical and refining industries, all the enterprises should be issued the pollutant permit before the end of 2018. In order to guide and regulate the application and issuance of emission permits for enterprises, the MEE issued the 'Technical Specification for Application and Issuance of Pollutant Permit in Petroleum Industry' (HJ 853-2017). At present, 34.46% of China's petrochemical and refining enterprises have completed the application of pollutant permit in the application platform. The VOC annual permit emission limit has implemented in three VOCs sources, i.e. Leakage of equipment static and dynamic sealing points, Volatilization loss of organic liquids storage and blending process and Volatilization loss of organic liquids loading and unloading process.

No pollutant emissions will be allowed without air pollution permits. One enterprise can held one permit. By implementing the pollutant permit system, it will change the phenomenon of disorderly discharge of enterprises in China, and promote the continuous improvement of environmental quality. The content of the permit license is not only to control the concentration of permitted emissions and the amount of permitted emissions, but more importantly, to control pollution prevention and control measures, finally to reduce pollutants.

Overall, from the review of the policies on the control of VOCs and/or photochemical smog, the stricter and aggressive measures help to alleviate air pollutants. Even though the atmospheric pollution is likely to be persistent because of on-going urbanization and industrialization, it is possible that the air quality can be gradually improved if appropriate measures are undertaken.

5 Problems and Suggestions

The Chinese government began to attach great importance to controlling VOCs emissions, regarding VOCs as a top priority among atmospheric pollutants for effectively improving air quality. Even though laws, regulations, standards and measures to control VOCs have been introduced in recent years, they still have problems and need to be further strengthened and standardized.

5.1 VOCs Monitoring Methods

Some standards put forward regulations and requirements to VOCs monitoring. However, the VOCs monitoring indicators and the monitoring technology methods are not clear or uniform. The characterization of VOCs is not uniform, some are based on the sum of characteristic pollutants, and some are based on NMHC. While in practice, VOCs are generally collected by solid adsorption or sampling bags, and analyzed by gas chromatography, gas chromatography/mass spectrometry, high performance liquid chromatography (HPLC) analysis, and a series of portable devices developed on this basis, such as TVA2020C, PHX-21. In addition, the pollutant permit system requires enterprises to implement self-monitoring. Now the self-monitoring of the pollutant permit has designed nearly 610,000 monitoring devices, while there are only 41077 related to the petrochemical and refining industries. Besides, the self-monitoring programs in enterprises are still based on manual monitoring, and the proportion of automatic monitoring is only 12.80%.

Therefore, it is necessary to formula and improve the VOCs monitoring standard system, establish scientific VOCs monitoring methods, and carry out research on petrochemical and industries industry monitoring method standards. In addition to monitor the conventional chimneys, leak detection standards for the devices (production devices, non-organized gas collection devices, etc.) should also be considered. Secondly, to improve monitoring capabilities, establish on-line monitoring networks for VOCs (using infrared instruments), and increase efforts to research portable monitoring devices and on-line monitoring devices. Finally, VOCs monitoring and big data platform are built to achieve standardized, precise, stable and reliable VOCs monitoring data.

5.2 VOCs Emission Inventory

The VOCs emission inventory is an indispensable parameter for the formulation of VOCs control measures. Compared with foreign countries, China's VOCs emission inventory research started late and is still on the beginning stage. There is no uniform and standardized emission source classification system, and there are few special studies on various types of emission sources. The petrochemical and refining industries have a poor level of refinement of emission source inventory, sluggish data updates, and non-comprehensive coverage. Besides, it lacks localized VOCs emission factors. The currently used emission factors are generally revised empirical emission factors that do not reflect China's actual situation. And the statistics of the activity level of the emission source sector is incomplete and lacks continuity.

Therefore, it is extremely necessary to establish an updated emission inventory that covers the spectrum of each pollution source in petrochemical and refining industries as much as possible. In addition, based on the actual situation in China, China's exemptions lists and the corresponding industry exemptions lists should be formulated. On the basis of the definition of photochemical reactivity, poisonous and harmful atmospheric pollutants shall be set up synchronously to control toxic and hazardous characteristics of VOCs. It is recommended to quote the list of HAPs formulated by the US EPA, and gradually improve the list of toxic and harmful air pollutants in China according to national conditions.

5.3 VOCs Control Technologies

The components and volumes of VOCs emitted from different processes in petrochemical and refining industries are different, and the adapted technologies are different. At present, the effect of VOCs treatment technology is different, and the non-methane hydrocarbon (NMHC) treatment efficiency is quite low. Moreover, the economic cost of some technologies is high, and many VOCs processing facilities cannot operate stably, continuously, and reliably.

Therefore, when selecting VOCs control technologies and facilities, enterprises should consider the characteristics of VOCs, as well as comprehensive consideration of various technologies' performance indicators, construction and operating costs, and implementation of emission standards. If necessary, two or more processes combination and multi-stage treatment can be used to achieve compliance with the emission standards. Special research should be conducted on VOCs in petrochemical and refining industries. Measures like continuous research and innovate treatment technologies, development relevant alternatives and improve existing equipment or processes are need to take effort to reduce VOCs emissions.

6 Conclusion

With the rapid progress of urbanization and industrialization in China, atmospheric pollution has become more and more serious. Currently, the type of atmospheric pollution is no longer a simple pollution, but a regional and complex pollution with extensive secondary pollutants. As an important precursor to cause secondary pollution, VOCs have attracted attention from countries in the world. Because of the huge VOCs emissions, wide range of sources in petrochemical and refining industries, accurately understanding the current status of VOCs emissions in the petrochemical and refining industries in China and the status of VOCs management and control are important prerequisites to improve China's ambient air quality. It is important for China's future comprehensive air pollution control decisions. Therefore, this article mainly discusses the status of China's atmospheric environment, the current status of VOCs emissions from petrochemical and refining industries, and the VOCs control measures.

China's management and control of VOCs in petrochemical and refining industries is still in its infancy. The future prevention and control of VOCs in petrochemical and refining industries remains an arduous task. In response to the current problems in the control of VOCs, some countermeasures and suggestions are proposed:

1) Formula and improve the VOCs monitoring standard system as soon as possible, and determine the characterization methods that comply with the VOCs in petrochemical and refining industries. At the same time improve the monitoring capabilities and the accuracy of online monitoring devices.

2) Accelerate the research on emission inventory and localization emission factors of VOCs in petrochemical and refining industries, establish an exemption list and a list of toxic and hazardous

pollutants, and provide technical support for the active implementation of emission inventory and source resolution of VOCs in the petrochemical industry;

3) Research and development best VOCs pollution prevention technologies for petrochemical and refining industries. Comprehensive consideration should be given when selecting VOCs treatment technologies. When necessary, a variety of technologies can be used together to achieve the pollutants meet the standards.

In the long-term perspective, the VOCs management and control model need to carry out work from the policy level, technical level, and management level. There are requirements and guidance in terms of policies, and they are technically supported and guaranteed in terms of technologies. There are breakthroughs and improvements in terms of management. Only by that, the VOCs pollution problem can effectively solved and the ambient air quality can be improved.

Acknowledgements

The work is supported by the International Cooperation Program for Innovative Talents of the China Scholarship Council (liujinwei [2016] 7988). Thanks for the funding.

References

1. Derwent R G, Jenkin M E, Saunders S M, et al. Photochemical ozone formation in north west Europe and its control[J]. Atmospheric Environment, 2003, 37(14):1983-1991.
2. Wu R, Xie S. Spatial Distribution of Ozone Formation in China Derived from Emissions of Speciated Volatile Organic Compounds[J]. Environmental Science & Technology, 2017, 51(5):2574.
3. Middleton P, Stockwell W R, Carter W P L. Aggregation and analysis of volatile organic compound emissions for regional modeling[J]. Atmospheric Environment Part A General Topics, 1990, 24(5):1107-1133.
4. Crutzen P J, Lawrence M G, Pöschl U. On the background photochemistry of tropospheric ozone[J]. Tellus Series B-chemical & Physical Meteorology, 1999, 51(1):123-146.
5. Klimont Z, Streets D G, Gupta S, et al. Anthropogenic emissions of non-methane volatile organic compounds in China[J]. Atmospheric Environment, 2002, 36(8):1309-1322.
6. Wei W, Wang S, Hao J, et al. Projection of anthropogenic volatile organic compounds (VOCs) emissions in China for the period 2010–2020[J]. Atmospheric Environment, 2011, 45(38): 6863-6871.
7. Saikawa E, Kim H, Zhong M, et al. Comparison of emissions inventories of anthropogenic air pollutants and greenhouse gases in China[J]. Atmospheric Chemistry & Physics, 2017, 17(10):1-41.
8. Huang G, Brook R, Crippa M, et al. Speciation of anthropogenic emissions of non-methane volatile organic compounds: a global gridded data set for 1970-2012[J]. Atmospheric Chemistry & Physics, 2017, 17(12):1-36.
9. Zheng J, Zheng Z, Yu Y, et al. Temporal, spatial characteristics and uncertainty of biogenic VOC emissions in the Pearl River Delta region, China[J]. Atmospheric Environment, 2010, 44(16):1960-1969.
10. Ohara T, Akimoto H, Kurokawa J, et al. An Asian emission inventory of anthropogenic emission sources for the period 1980–2020[J]. Atmospheric Chemistry & Physics, 2007, 7(16):6843-6902.
11. Xia L, Cai C, Zhu B, et al. Source apportionment of VOCs in a suburb of Nanjing, China, in autumn and winter[J]. Journal of Atmospheric Chemistry, 2014, 71(3):175-193.

12. Ying L, Min S, Lu S H, et al. Source apportionment of ambient volatile organic compounds in the Pearl River Delta, China: Part II.[J]. Atmospheric Environment, 2008, 42(25):6261-6274.
13. Rumei L I, Yuanyuan W U, Peng L, et al. Characteristics and sources apportionment of ambient volatile organic compounds(VOCs) in summer in Shuozhou[J]. Environmental Chemistry, 2017.
15. Han L, Wang X R, He M, et al. Inventory and environmental impact of VOCs emission from the typical anthropogenic sources in Sichuan province[J]. Environmental Science, 2013, 34(12):4535-4542.
16. Wallace L A, Pellizzari E D, Hartwell T D, et al. The influence of personal activities on exposure to volatile organic compounds.[J]. Environmental Research, 1989, 50(1):37.
17. Zhou J, You Y, Bai Z, et al. Health risk assessment of personal inhalation exposure to volatile organic compounds in Tianjin, China.[J]. Science of the Total Environment, 2011, 409(3):452.
18. Lin L, Jun L U, Ying-Ge M A, et al. Reviews of VOCs Control at Home and Abroad[J]. Administration & Technique of Environmental Monitoring, 2011.
19. Wang H L, Zhang G N, Nei L, et al. Study on control and management for industrial volatile organic compounds (VOCs) in China.[J]. Environmental Science, 2011, 32(12):3462.
21. Guo S, Tong L, Zhou X, et al. VOCs Emission Control in Petrochemical Industry[J]. Environmental Protection of Chemical Industry, 2014.
22. Zhang G N, Hao Z P, Jiang M, et al. Study on foreign regulations and standards of stationary sources VOCs emission control.[J]. Environmental Science, 2011, 32(12):3501.
23. Louie P K K, Ho J W K, Tsang R C W, et al. VOCs and OVOCs distribution and control policy implications in Pearl River Delta region, China[J]. Atmospheric Environment, 2013, 76(S1):125-135.
24. Zhang X, Xue Z, Sun X, et al. Review on Control of Ambient Volatile Organic Compounds (VOCs) and Suggestions in China[J]. Environmental Science & Management, 2014.
25. Khan F I, Ghoshal A K. Removal of Volatile Organic Compounds from polluted air[J]. Journal of Loss Prevention in the Process Industries, 2000, 13(6):527-545.
26. Guo H, Ling Z H, Cheng H R, et al. Tropospheric volatile organic compounds in China[J]. Science of the Total Environment, 2017, 574: 1021-1043.

513535 A Look inside a Chemical Plant Waste Water Treatment Unit

Abstract:

There are a wide variety of pieces of equipment that compose waste water treatment inside of chemical plants. In this presentation, I will discuss the process flow of the waste water treatment plant for the Syngenta St. Gabriel chemical plant to help young engineers understand the complexities of such a system and the current commercially available systems we use to treat waste water to meet government regulations. We have a pre-treatment step, carbon adsorption, and then a biological unit that help to treat the water before it can be sent out to the Mississippi River.

516714 Non-Ideal Two-Phase Flow on Large Diameter Trays: The Stagnant Regions Model versus FRI Experiment

Abstract:

Forty six years ago, the Stagnant Regions Model (SRM) was published by Porter, Lockett and Lim in which the impact of severe flow maldistributions on tray efficiency for large diameter trays was investigated. The SRM involves two zones to characterize the liquid flow pattern on a single-pass tray: an active liquid channeling zone between the downcomers and retrograde/stagnant zones at the sides of the tray. It was found that for large diameter sieve trays, a loss in tray efficiency was reported due to liquid channeling and associated stagnant zones. The SRM theory was supported experimentally years later in the Aston University 1.22 and 2.44 m air-water simulator columns (Porter et. al., 1987, 1992 and 1995). At the same time the SRM was published, FRI studies on maldistributed liquid flows on 1-pass sieve trays using hydrocarbon systems in the 2.44 m FRI Low Pressure column showed no loss in tray efficiency both with and without flow control mechanisms (Bell, 1972; Yanagi and Scott, 1973).

The purpose of this paper is to discuss tray geometry parameters employed in the SRM to explain why the model predicted no loss in tray efficiency on the FRI 2.44 m diameter 1-pass sieve tray. One important parameter to be investigated is Width-of-the-Mixing-Zone (WMZ) at the sides of the tray. Arguments will be supported with FRI film footage of liquid distribution both without and with flow control mechanisms.

512752 Evolution of Random Packing Performance with a Focus on Geometry

Abstract:

The geometries of random packings have evolved significantly over the years to offer a broader range of hydraulic and mass-transfer performance. Packings were originally made from local materials, but new generation packings with engineered geometries have been designed and manufactured since the 1950â €™s to significantly increase specific surface area with improved pressure drop. This presentation will investigate how packing geometry affects form drag and friction drag with the evolution of packings over time, and how they are addressed with fundamental correlations.

Key Words: Random packing, Packed Columns, Pressure Drop, Correlations

*Corresponding author. E-mail: ken.mccarley@fri.org , Phone: 1-405-385-0354, Fax: 1-405-385-0357

512806 Computational Fluid Dynamic (CFD) Simulations of Multiphase Flow through Representative Elementary Units (REU) of Structured Packing

Abstract:

Computational Fluid Dynamic (CFD) Simulations of Multiphase Flow Through Representative Elementary Units (REU) of Structured Packing

Trien (Mikey) Phan and R. Bruce Eldridge

James R. Fair Process Science and Technology Center

The University of Texas at Austin

Current model predictions of the hydraulic behavior and separation performance of structured packings rely on semi-empirical correlations whose ranges of viability are limited to the packings and operating conditions used during correlation development. A more fundamental understanding of the momentum transfer inside structured packings is crucial for improving model predictions of the hydraulic properties of structured packing, which in turn can lead to optimized designs of mass transfer column internals. To this end, we utilized Computational Fluid Dynamics (CFD) to simulate gas-liquid flows inside simplified Representative Elementary Units (REU) of structured packing. Using Siemensâ€™ STAR-CCM+ CFD software, we investigated the impact of packing geometry, such as corrugation angle and flow channel dimensions, as well as fluid properties, such as surface tension and contact angle, on the hydraulic performance of an REU. The modeling results yielded an optimum geometry for the hydraulic performance of structured packing.

513259 Evidence for the Need of Probability Density Function Representations of Liquid-Side Mass Transfer Rate Coefficients for Absorption Modeling in Structured Packings

Abstract:

Gas-liquid mass transfer parameters of structured packings are generally determined from global absorption or desorption measurements over some bed height, sometimes with treatment of entrance and outlet effects. Correlations of these global parameters are then developed relative to the overall flow conditions and fluid properties and often take the classical form

Eq. 1

where and stands either for the gas phase, G, or the liquid phase, L. These equations are subsequently applied in models for sizing large absorption or distillation columns using local average flow conditions and fluid properties. Rejl et al. [1] have recently shown that the dependencies to the flow conditions in such correlations, while allowing to match global features, such as HETP, do not accurately reproduce axial concentration profiles. They propose a so-called "profile method" to obtain better functions relative to flow conditions. In this work we present other experimental evidence that indicates that correlations of mass transfer parameters derived from a global perspective may be insufficient to properly describe the details of the processes required for column sizing.

We consider the case of physical absorption or desorption. Based on the double-film model the global overall gas-side mass transfer resistance ($1/K_Ga$) is mathematically the sum of the global gas-side ($1/k_Ga$) and liquid –side (m/k_La) resistances:

$1/K_Ga=1/k_Ga+m/k_La$ or $R_{G,ov} = R_G+mR_L$ Eq. 2

where m is the slope of the equilibrium curve. This equation is valid if all liquid-gas interfacial areas are subjected to the same mass transfer rates. However, in many situations such is not the case. Haroun et al. [2] have done CFD calculations of film flow over a packing structure and indicate that, in coherence with the Higbie contact time surface renewal model, the liquid side mass transfer coefficient decreases inversely with the distance or time from the renewal point. Recent Xray tomographic imaging [3, 4] and CFD simulations [5,6] indicate the existence of flow heterogeneities leading to rivulets formation and unwetted (dry) regions, even with hydrocarbon liquids that have better wetting characteristics than water.

Now we further consider a case where the global or average mR_L is secondary to R_G . If there exists a local region in the packing where the liquid film velocity is low then it will likely be characterized by a longer renewal time relative to the mean. This will lead to a locally higher R_L that will tend to reduce the overall K_G and hence the mass transfer rate. In contrast, if there is a region with very low R_L, say along a rivulet, then there will be no impact on the overall mass transfer rate since the liquid side resistance is already globally low. The opposite will hold for a case where global liquid side resistance is large. Therefore local flow heterogeneities will, under certain conditions, alter the true mass transfer rate relative to one predicted by liquid and gas-side mass transfer parameters determined independently and on a global basis.

To illustrate this fact, we present experiments that have been performed with two different chemical systems to measure the global k_Ga of Mellapak 250.X packing in a 146 mm diameter column. One system is the fast reactive absorption of SO_2 diluted with air into a 2N solution of NaOH. This system is well known to be only gas side limited [9]. The second system is the physical absorption of ethanol or methanol diluted with air into water for which the contribution of liquid side resistance to the overall resistance is often small. So from a global perspective, with the reasonable assumption of negligible liquid side resistance, both systems should give the same relation for Sh_G as given in Eq. 1. Yet the values of Sh_G determined from SO_2 absorption are 50 % higher than from ethanol absorption (Figure 1). Based on global mass transfer parameters from the literature [7] , under our flow conditions, the contribution of liquid side resistance to the total does not exceed 10% at the highest gas flowrate. The 50% difference is well beyond the experimental uncertainties of the global k_L measurements. For example, it would be necessary to excessively reduce k_L by 4 to 5 times in order to make the data coherent. In order to help understand the different results between the two systems, further experiments were carried out in a 19 mm ID column over a wire gauze high effective area packing (Sulzer EX) at very low liquid loads. The liquid side resistance was measured independently and found to contribute by at most 5% to the overall resistance. In this situation the two chemical systems gave the same Sh_G values within 10%.

Figure 1. Comparison of Sh_G values obtained with different systems over Mellapak 250.X packing at a liquid load of 30 m/h. The values have been normalized relative to the one obtained with SO_2 absorption at $Re^{2/3}Sc^{1/3}=100$.

To explain these contrasting results we propose to consider the nature of the liquid film flow. In the wire gauze packing the liquid film spreads very uniformly by capillary action and likely stays very smooth. In the Mellapak 250.X packing the heterogeneities in the film flow are likely to lead to some form of distribution of local k_L or R_L values. The existence of k_L distributions has been discussed in the past by Danckwerts [8] in terms of a distribution of renewal times and how it impacts the expression for the global k_L. However no interaction of this distribution with the overall resistance to mass transfer was discussed. If the distribution is described by a probability density function, $f(k_L)$, then the fraction of effective area that is subjected to liquid side mass transfer coefficients of values between $k_L \pm dk_L/2$ is given by $f(k_L) \cdot dk_L$. The average liquid side mass transfer coefficient and overall resistance are then calculated as

Eq. 3

Eq. 4

The distribution function should yield $= k_L$ and, based on the experimental results, $/R_G = 2/3$. As an exercise, we found that a log-normal probability density function on k_L was able to satisfy these conditions whereas simple normal distribution function could not. The true nature of the distribution should be the topic of future work.

In conclusion the local liquid flow heterogeneities that can exist in random and structured packings possibly produce a distribution of local liquid-side resistances. The use of a global liquid-side mass transfer coefficients may then lead to erroneous sizing of physical absorption or distillation columns in situations where both liquid and gas-side resistances contribute to the overall mass transfer resistance. Under reactive absorption the k_L distribution will furthermore affect the local Hatta number (Ha) and lead to a distribution of mass transfer regimes. Evaluation of the consequences these distributions should be the topic of future work. Last we conclude that physical absorption methods are not suited for measuring the global k_Ga in random or structured packing when liquid flow heterogeneities exist, instead, reactive absorption of SO_2 in NaOH should be used.

[1] F. J. Rejl, L. Valenz, and V. Linek, "Profile Method" for the Measurement of kLa and kVa in Distillation Columns. Validation of Rate-Based Distillation Models Using Concentration Profiles Measured along the Column , *Ind. Eng. Chem. Res.*, 49 (9) (2010) 4383–4398

[2] Haroun, Y., Raynal, L., Legendre, D., Mass transfer and liquid hold-up determination in structured packing by CFD. *Chem. Eng. Sci.,* 75 (2012) 342–348.

[3] S. Schug and W. Arlt, Imaging of Fluid Dynamics in a Structured Packing Using X-ray Computed

Tomography, *Chemical Engineering and Technology* 39 (2016) 1561–1569.

[4] A. Janzen, J.Steube, S.Aferka, E.Y.Kenig, M.Crine, P.Marchot, D.Toye, Investigation of liquid flow morphology inside a structured packing using X-ray tomography. *Chemical Engineering Science* 102 (2013) 451–460

[5] Y. Haroun, L. Raynal, P. Alix, Prediction of effective area and liquid hold-up in structured packings by CFD, *Chemical Engineering Research and Design,* 92 (2014) 2247–2254.

[6] Bing Dong, X.G. Yuan , K.T. Yu, Determination of liquid mass-transfer coefficients for the absorption of CO_2 in alkaline aqueous solutions in structured packing using numerical simulations , *Chemical Engineering Research and Design* 124 (2017) 238–251.

[7] C. Wang, D. Song, F. A. Seibert, G. T. Rochelle, Dimensionless Models for Predicting the Effective Area, Liquid-Film, I*nd. Eng. Chem. Res.*, 55 (2016) 5373–5384. .

[8] Danckwerts, P.V., 1970. *Gas-Liquid Reactions*. McGraw-Hill, New York.

[9] L. Hegely, J. Roesler, P. Alix, D. Rouzineau, M. Meyer, Absorption methods for the determination of mass transfer parameters of packing internals: A literature review, *AIChE Journal* 63 - 9 (2017) 3246–3275.

Distillation Control – Practical Aspects

Sigifredo (Siegfried) Nino, P. Eng.

Summa Control Solutions Inc.

Montreal, QC, CANADA

SNino@SummaControlSolutions.com

Presented at the 2018 AIChE Spring Meeting, Advances in Distillation I, Orlando, Florida, April 22nd – 26th, 2018

UNPUBLISHED

Distillation Control – Practical Aspects

Siegfried Nino, P.E. - SNino@SummaControlSolutions.com

Process Control Consultant, Summa Control Solutions Inc., Montreal, Canada

One of the most important aspects of implementing, commissioning or troubleshooting any distillation control strategy, resides in the recognition that modifications to existing equipment are improbable. This paper illustrates several cases where changes to original control strategy designs were made to be able to complete a commissioning or resolve long-standing issues relating to distillation control.

Process control makes a significant contribution of to the improvement of plant operation stability. Improving a unit's stability may appear, at first, to have no benefits to the operation or to the optimization of a chemical processing facility. Yet stability enhances the clarity in the landscape of the process' variables and as a result, opportunities for improvement emerge as oscillations are tamed. Included in this writing are relevant and practical tips to illustrate the benefits of this practice.

BENZENE TOWER - TEMPERATURE DIFFERENCE VS. COMPOSITION CONCAVE RELATIONSHIP

Background

This tower is part of a Benzene-Toluene-Xylene (BTX) train at a refinery. Its purpose is to separate the benzene from the toluene. Both products must remain at high purity. The column has 54 trays counted from the top. The feed enters at tray 31 and the benzene product is withdrawn as sidestream from tray 6. The lighter material is removed from the overhead condenser and steam is used to provide the heat input while air coolers are used to condense the overhead vapour. Most of the feed leaves through the bottom product with 0.05% BZ impurity in the Toluene stream specification.

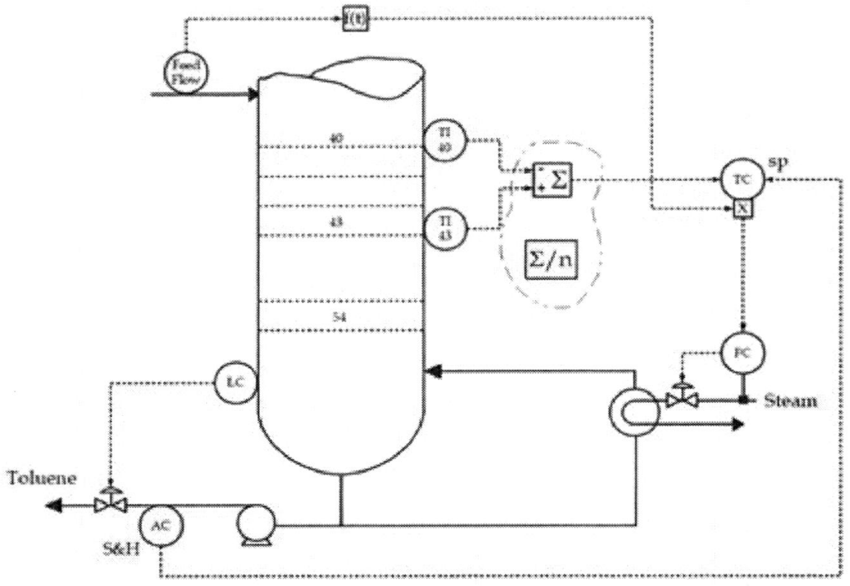

Figure 1 - Benzene Tower - Bottoms detail

Two differential temperature measurements were originally used for double-ended composition material-balance control. The top differential temperature (ΔT), is calculated between plates #2 and #14 as the surrogate of toluene impurities in the benzene product. The dual-end composition control of this high-purity distillation column was completed using a bottom section ΔT, calculated by using sensors located three trays apart (Tray #40 and Tray #43). These temperature sensors were RTD (Pt 100 DIN 43760-1968) ranged 0-1600 °F, with an accuracy no better than ±0.25% of span or ±4 °F, for a differential temperature in the range of 4°F, causing a sizeable degree of uncertainty in this measurement.

Figure 2 shows the test performed to qualify and quantify the use of the differential temperature as benzene in the bottom product surrogate. The bottom composition control implemented is the double cascade array shown in Figure 1, where the outmost PID controller operates in a Sample-And-Hold mode synchronized with the "new sample" signal from the bottom product in-line analyzer, setting the setpoint to the differential temperature controller. In turn, this controller sets the steam supply required to keep the tower boilup. In normal operation, a fall in differential temperature is due to the reduction of the light key in the bottom, so the controller reduces boilup, and then both differential temperature and bottom composition will rise to the desired values.

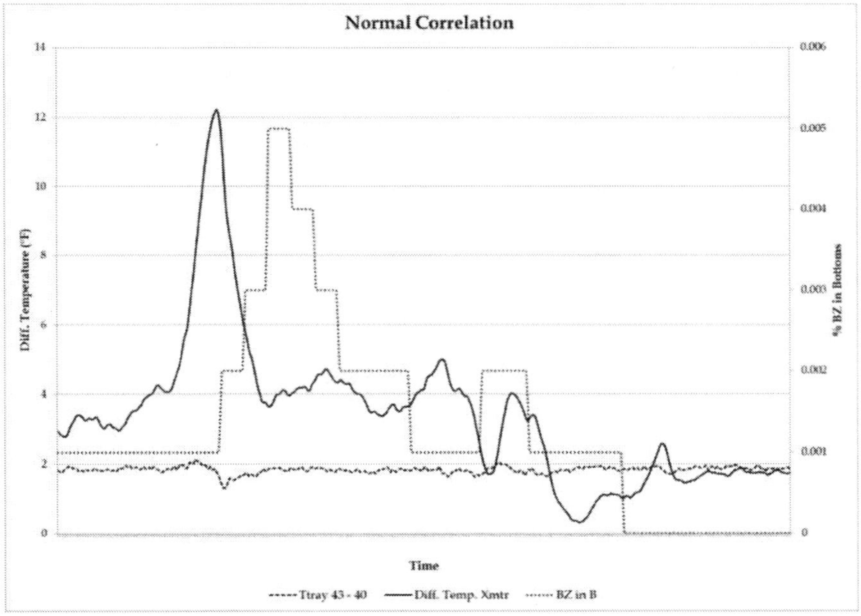

Figure 2- Benzene Tower - Differential Temperature Test

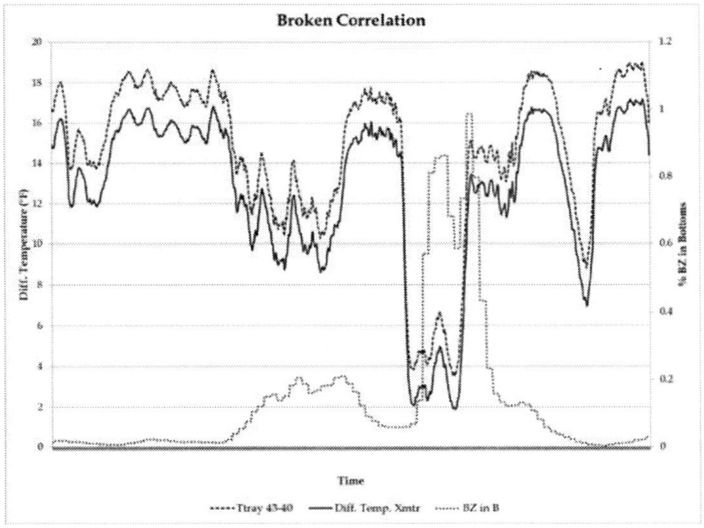

Figure 3 - Benzene Tower - Breaking ΔT and Composition relationship

Using a similar analysis designed by Webber (1959), who used the McCabe-Thiele diagram, it was found in the exercise that two orders of magnitude difference in product composition, will cause the Tray40-Tray43 versus Bottom impurities to follow the concave curve in Figure 4. The operation on the left-hand side of the curve means that if the ΔT setpoint equals 10°F and the measurement is 12°F, the controller will need to increase the steam in order to reduce the temperature difference. However, the ΔT could also be at 12°F for a composition of 0.17 Benzene in the bottom product, the controller will then try to reduce the ΔT by cutting the steam to bring ΔT back to 10°F. This will, however, cause the composition to be even farther away from the setpoint, as more impurities will reach the bottom of the tower.

For the sake of completing the topic, it is worth mentioning that Kister (1990) suggested yet another cause for the concavity of the composition versus Differential temperature. He stated that it can also be related to changes in tower pressure, even though it has been said that a differential temperature in a distillation column is immune to pressure variations.

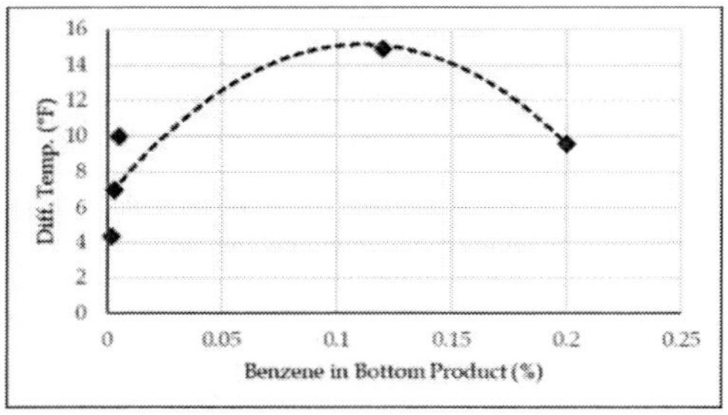

Figure 4 - Composition vs. dT concavity

Solution

The solution to the inversion in the relationship was overcome by using the temperature sensor's average, as indicated by the cloud in Figure 1.

A relevant issue related to the application is illustrated in Figure 4. Note that at very low composition the temperature difference could be in the order of 4°F. Although the original RTD's where replaced with high precision sensors, the measurement of the differential was done in the DCS by subtracting the two individual temperatures, causing the ΔT to become negative at times, a physical impossibility from the process' point of view. The recommendation for this type of applications is to have the two sensors go into a temperature difference measuring device instead.

The direct benefit of this material-balance dual-composition control in a high-purity distillation application was a 35% reduction in energy consumption.

TOLUENE TOWER - THE IMPOSSIBLE CONTROL TRAY TEMPERATURE

Background

This toluene tower is downstream of the benzene tower in a BTX train at a Petrochemical facility. These two towers are thermally integrated; the toluene tower's overhead vapour latent heat supplies the energy input to the benzene tower. Dual-composition control was implemented using Structured Advanced Regulatory Controls.

The tower has 72 trays counting from the bottom. It operates with a flooded reflux drum, most of the product leaves through the top product, and the bottom product is mostly xylene with 1.1% Toluene in the bottom product specification.

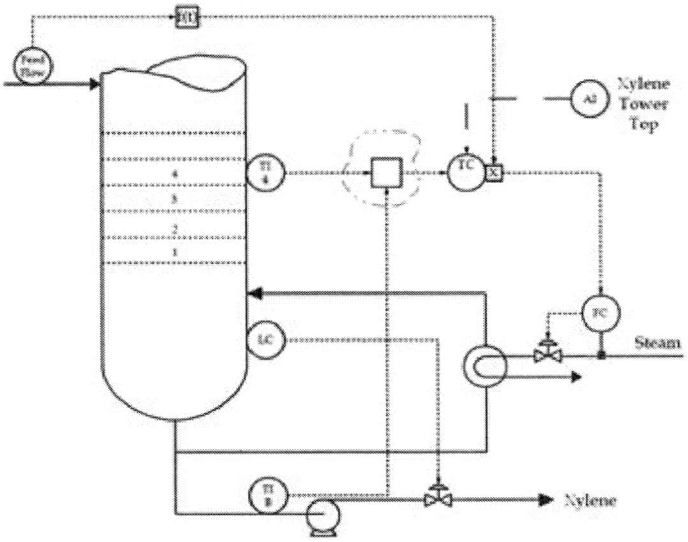

Figure 5 - Toluene Tower - Bottom Composition Control

An attempt to control the toluene impurities in the stream towards the xylene tower had been made by using a model-based composition controller that used a Near-Infrared (NIR) analyzer to reset the toluene tower tray #4 temperature surrogate, which in turn manipulated the Boilup to Feed ratio, as shown in Figure 5.

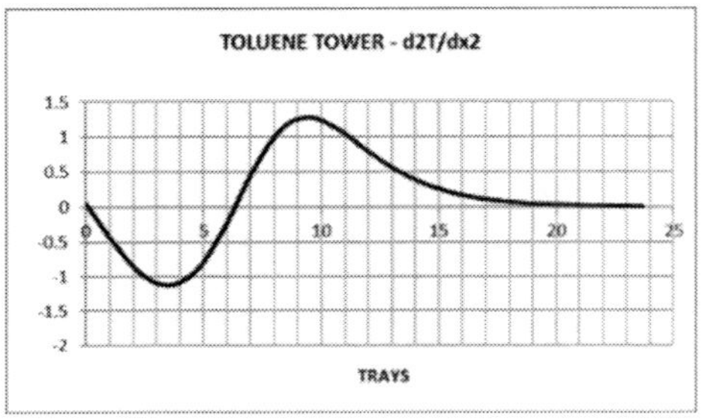

Figure 6 - Toluene Tower - Second Derivative of Temp. Profile

The controller didn't really work. In steady state operation, it would work fine, but every time there was an enthalpy disturbance, the composition would follow such extreme excursions that it would take operators several hours, typically a whole shift, to return the composition to specification. This problem didn't stem from the use of a Model-Based controller. It came from the whims of an uncontrollable variable in the inner loop of the cascade. This is to say that Model-Based controllers are not magical; they will work fine when they are placed in the right position in the control hierarchy.

Solution

There were two causes in this problem: Control interaction in the benzene tower, and control plate temperature hypersensitivity. Both are described below followed by the implemented solution.

The paper that presented this case (Nino, 2011) showed that by using the relative gain analysis, the benzene tower's original conventional dual composition configuration control LV (Reflux set to control top impurities and Boilup to control bottom impurities), resulted in severe interaction causing great disturbances that were propagated to the toluene column through the heat integration system. As detailed in the aforementioned paper, this problem was successfully resolved by implementing material-balance dual-composition control in both towers.

The investigation into the second source of the problem started by taking the tower temperature profile, available from a tray-to-tray simulation, to confirm that the temperature sensor was properly placed in the tower. Figure 6 shows the plot of the second derivative of the tower temperature profile and confirms that the location of the temperature sensor at tray #4 corresponds to a point of maximum sensitivity in the tray-to-tray compositions changes. In the actual application, however, all the attempts to control this tray temperature were unsuccessful. The clarification to this riddle came from the plotting of Tray Temperature versus bottoms product Impurity, illustrated in

Figure 7. It became clear that there are hypersensitivity and even relation reversal, that lead to conditional stability of the controls.

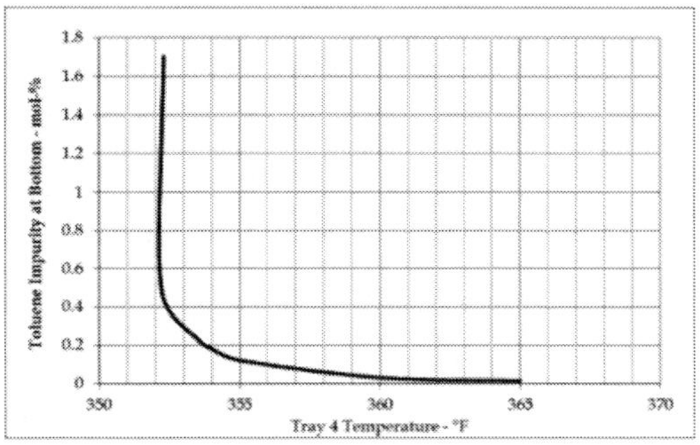

Figure 7 - Toluene Tower - Bottom Temperature vs. Composition

An alternate temperature surrogate was found in the bottom product temperature, indicated as TI-B in Figure 5. Testing proved that there is a repeatable relationship with the desired composition. The controller was commissioned for the new configuration and it has been working for more than half a decade.

As mentioned before, a NIR analyzer sampling the Xylene tower's top product was being used for resetting the temperature surrogate. However, this would introduce unwanted disturbances to the control, due to the low analyzer's signal-to-noise ratio. The facility then chose to have the setpoint done by the operator and based on daily lab testing.

The application of this strategy resulted in over $1 MM savings in energy usage and reduction of product recycle. See Nino (2011) for details.

DEISOBUTANIZER - TRIPLE SPLIT RANGE

Background

This sidestream alkylation Deisobutanizer (DIB) operates at a recently commissioned oil refinery. The issues: The original overhead accumulator level control was done through the sequencing of three valves, and the technology of the flowmeters selected for the application make them unreliable for this application. This is a work in progress as a material balance control will be needed to optimize the tower operation.

The overhead accumulator is the liquid storage supply for the reflux and the column top product, isobutane in this case. There are two different destinations for the isobutane, alkylation plant and storage. The first one being the preferred route, and the second the excess production being sent to tanks. The drum level control strategy was designed and configured to be a triple split-range scheme: 0-33%, 33-66%, and 66-100%. The first part of the split was supposed to keep a constant flow to the alky plant once the controller output exceeded 33%, and the remaining 66% controller output should have taken care of any variability in the reflux drum inventory.

The commissioning of the DIB was done before that of the alky plant, unveiling the inability of the strategy to keep the overhead drum under control. The operators ended up operating the loop in manual, which implied frequent moves intended to prevent the drum from flooding or emptying. The abnormal situation did not change even after the start-up of the Alky Plant. The strategy could not work and the operators were continuously changing the controller output to keep the drum "in control".

Solution

In a process where there is a pump attached to a storage, the average level responds as an integrator. Thus, the process becomes non-self-regulated and in the absence of an able controller, it will require constant operator attention.

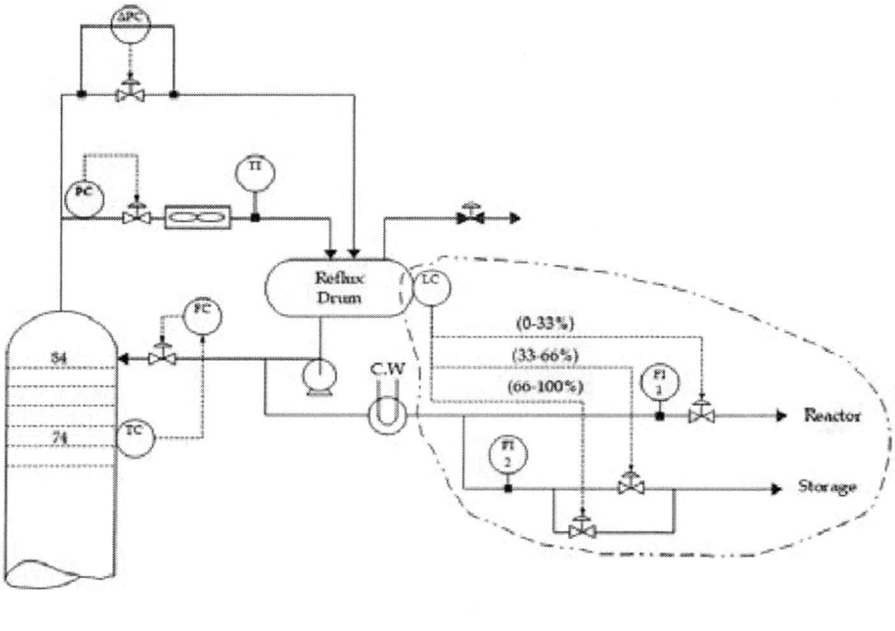

Figure 8 – Alkylation Deisobutanizer

The problem was resolved by assigning the first control valve to the stand-alone flow control loop, FC-1 in Figure 9. This ensures a constant feed flow to the alky plant. The other two valves, being the same size and having linear trim were assigned to be the manipulated variable for the level control loop, as shown in Figure 9. A second control loop would be the right solution to implement, but the operation of the second flowmeter

(FI-2 in Figure 9) is not reliable because "Vortex flowmeters require turbulent flow to operate and will cease to read as the fluid transitions from the turbulent flow regime to the transitional or laminar flow regime. This is called the "low-flow cut-off" point of the meter, and flow rate measurement below this point is not possible" (Mc. Millan et al,

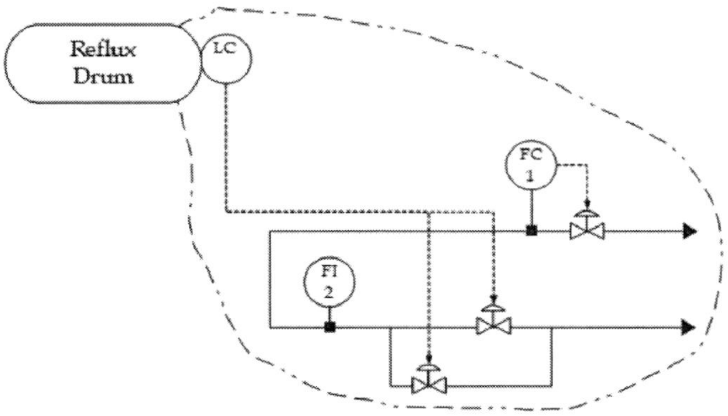

Figure 9 - Deisobutanizer - Reflux drum level control

2013). Whilst the flow rate to the alky plant is maintained over the minimum "readable", that is not always the case for the flow to storage tanks.

As mentioned above, a better solution could be implemented by using a material balance control to coordinate the flow loops, which would require reliable flowmeter operation.

Again, a lesson that should have been learnt a long time ago: The open-loop unstable systems are operationally critical, and although operators can take care of them in exceptional circumstances, when it becomes the norm, it's bound to cause more noticeable disturbances to the unit or the plant operation.

The benefits in this case include reducing operator attention and time spent in this unit, freeing the Operator to maintain awareness of all units assigned to him/her.

OSCILLATING ATMOSPHERIC TOWER

Background

The crude unit at this refinery was subject to 50-60 minutes oscillation (Figure 11) thought to be originated in the fuel gas network. The oscillation was traced back to the crude tower. It was found out that in the crude unit original control strategies design, the tower pressure control, indicated with a hash background circle in Figure 10, was associated to a manipulated variable that had virtually no effect on the controlled variable.

A validation test performed in the unit proved that the chosen pressure controller was more related to maintaining a constant compressor suction pressure and that its effect on the crude tower pressure was insignificant.

Solution

Distillation column pressure is one of the most difficult distillation variables to control. The pressure in a distillation tower can be varied with Temperature, Volume or Mass, the first two are not really available so the only option available is the control the vapor inventory.

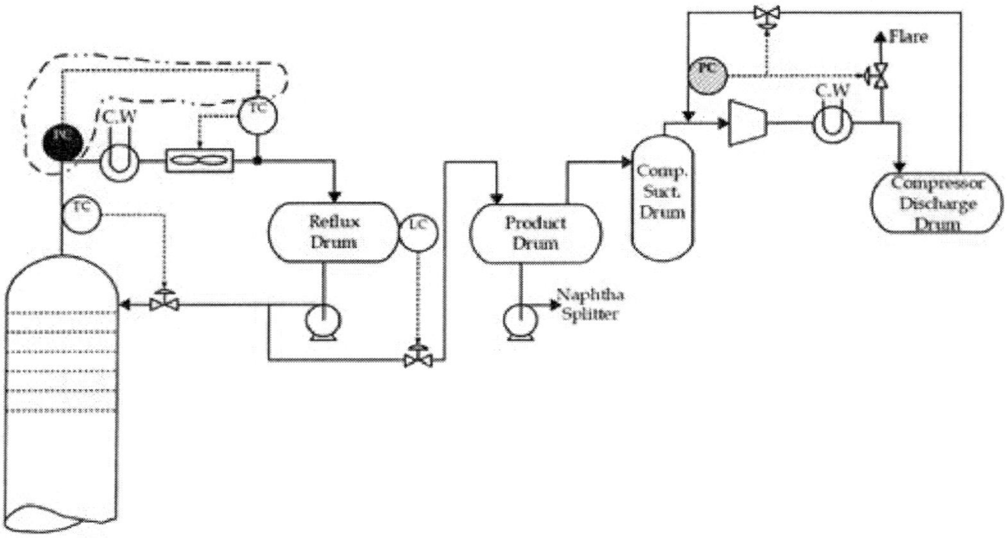

Figure 10 - Crude tower - Tower Pressure Control- (C.W. is Cooling Water)

The condenser cooling gave an additional degree of freedom to the system, and the relevant testing showed that the tower pressure would be successfully controlled by manipulating the Air Coolers' exit temperature in a cascade control with the pressure controller, as emphasized by the black background circle, on the left-hand side of Figure 10.

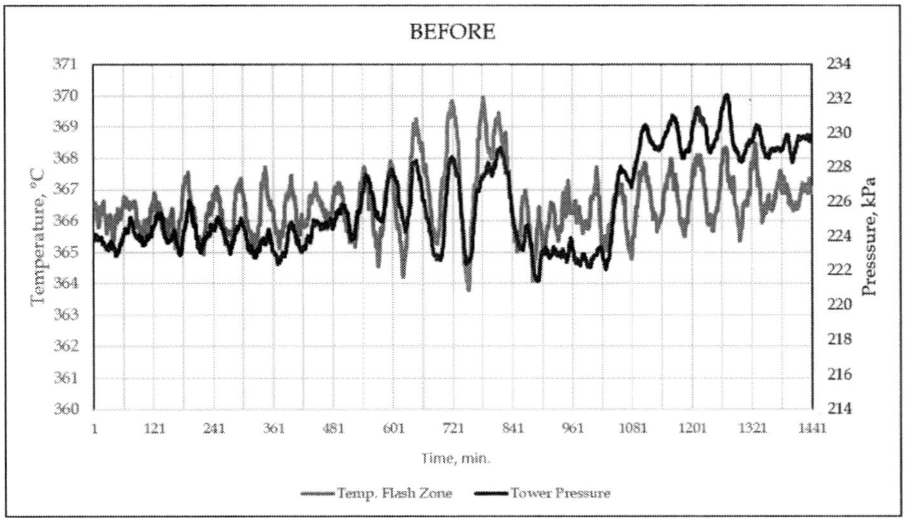

Figure 11- Crude Tower Oscillating Pressure – Before

After the commissioning of the new pressure controller, the entire tower temperature profile dramatically improved (Figure 12). It was much easier to separate the different cuts as the cuts overlapping was reduced.

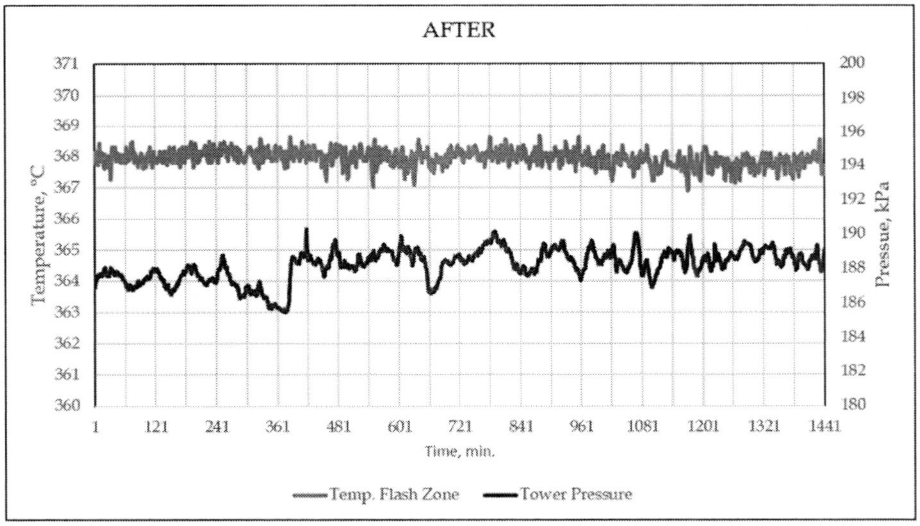

Figure 12 - Crude Tower Oscillating Pressure – After

This resulted in the pressure standard deviation reducing from 8 kPa to 0.5 kPa, i.e. sixteen (16) times reduction, and the peak-to-peak deviation was reduced by one order of magnitude, from 34 kPa to 3.4 kPa.

BIBLIOGRAPHY AND REFERENCES

[1] Chin, T. G. (1979) "Guide to distillation pressure controls", Hydrocarbon Processing, October

[2] Kister, H. Z. (1990) *Distillation Operation*, McGraw-Hill, Boston

[3] Nino, S. (2011) "Improve material balance in high-purity distillation control", Hydrocarbon Processing, November, p.83-88

[4] Nino, S. (2017) "Distillation Control: Still an Issue", in Kister Distillation Symposium 2017 proceedings

[5] Rademaker, O.; J. E. Rijnsdorp and A. Maarleveld (1975) *Dynamics and Control of Continuous Distillation Units*. Elsevier Scientific Publishing Company, Amsterdam. P. 375

[6] Shinskey, F. G. (1984) *Distillation Control: For Productivity and Energy Conservation*, 2nd Edition, McGraw-Hill

[7] Sloley, A. W. (2001) "Effectively Control Column Pressure", Chemical Engineering Progress, January

[8] Webber W. O. (1959) "Control by temperature difference?" Petroleum Refiner V.38, No.5

[9] McMillan G. K.; Vegas H. (2013) 101 Tips for a Successful Automation Career, ISA

514036 Packed Tower Separation Efficiency Problem Diagnosed Using Gamma Scanning Technology

Abstract:

Dow Chemical was experiencing separation efficiency problems in one of its product towers. Tower internals consisted of four beds of structured sheet metal packing. Process calculations showed the tower was operating with 40% less equilibrium stages than design. Liquid maldistribution was expected and gamma scans were completed. In addition to using the standard four path grid scan layout, two additional scan lines crossing through the center of the tower were done to increase the chances of detecting maldistribution.

Effort was made to quantify the liquid maldistribution based on the gamma scan results to serve as input into a maldistribution sensitivity study. The conclusion from this analysis was the estimated maldistribution was great enough to have a significant effect on the tower efficiency. Not able to discern the root cause of the liquid maldistribution at the time (e.g. distributor design, fouled distributors, installation problem, problem in the packing, etc.), new high quality liquid distributors were designed and installed.

After installation of new liquid distributors, the tower started up and the separation efficiency improved to 98% of the design stages. This presentation includes discussion and results focused on the gamma scans and what can be done with gamma scan data to help with troubleshooting liquid maldistribution.

512776 Thermodynamic Modeling for CO_2 Absorption in Aqueous Potassium Carbonate Solution with Electrolyte NRTL Model

Abstract:

Potassium carbonate has recently gained attention and is considered favorable for CO_2 removal due to numerous reasons such as low economic costs and energy consumption, low volatilization and degradation and less environmental impacts compared to amine solutions. Therefore, we develop a comprehensive thermodynamic model for the $K_2CO_3 - H_2O - CO_2$ system using the Electrolyte Nonrandom Two-liquid (eNRTL) activity coefficient model in conjunction with PC-SAFT equation of state. Vapor-liquid equilibrium (VLE), heat capacity, excess enthalpy, mean ionic activity coefficient and osmotic coefficient data are used to determine eNRTL temperature dependent binary interaction parameters for both $K_2CO_3 - H_2O$ binary and $K_2CO_3 - H_2O - CO_2$ ternary systems. The model accurately represents all the thermodynamic properties and covers a wide temperature range from 273.15 to 473.15 K and CO_2 loading from 0 to 1 The model can be merged with other amine models for mixed K_2CO_3 – amine solutions and the model can be used for comparison with other CO_2 capture chemical solutions. The model should be a useful tool in the process modeling, simulation and design of CO_2 capture processes involving hot, concentrated K_2CO_3 solutions at elevated temperatures and pressures.

512813 Single and Multi-Objective Optimizations Using Parallelized Process Simulators

Abstract:

Equation-oriented (EO) optimization methods are difficult to apply to rigorous thermodynamic methods, discontinuous variables, or systems where the number of equations is not constant. Despite these challenges, EO methods are still quite commonly used for process optimization, due in part to a perception that EO methods are faster than process simulation-based methods.

By integrating parallel processing techniques into modern process simulation and optimization, rapid results can be achieved comparable to the speed of more traditional EO modeling methods. These speeds can be attained even while using first-principle models, without risking validity of the model through too many simplifying assumptions. In this presentation, several process examples will be considered to explore the advantages of using simulators for optimization projects.

512830 Thermodynamic Modeling of the $LiCl+LiBr+LiNO_3+LiI+H_2O$ System with the Electrolyte NRTL Model

Abstract:

Thermodynamic models are essential to facilitate the advancement of process design, optimization, and operation range of absorption heat pumps and refrigeration units that utilize lithium salt systems. The lithium bromide-water solution is a popular working fluid due to its environmentally friendly, nonvolatile, and non-toxic properties; however, this solution has issues with corrosion and crystallization. One proposed solution is to modify the lithium bromide-water working solution by adding anticrystallization and anticorrosion agents such as lithium chloride, lithium iodide, and lithium nitrate. The objective of the present work is to develop a comprehensive thermodynamic representation of the $LiCl+LiBr+LiNO_3+LiI+H_2O$ quinary solution by implementing the symmetric electrolyte NRTL (eNRTL) activity coefficient model. This solubility model correlates the nonideality of the solution with the composition dependency by utilizing two binary interaction parameters for each of the electrolyte-electrolyte and molecule-electrolyte pairs in the system. Each parameter is determined by applying up to three temperature dependent coefficients and then regressed from experimental data. This allows the parameters to establish the accuracy of the model for concentrations up to saturation and temperatures between 273.15-473.15 K.

514059 Image Classification in Manufacturing Analytics: Improving a Pellet Classification System with Deep Neural Networks

Abstract:

Manufacturing analytics is of paramount importance in many plants today, and its significance increases in the current context of big data analytics and industry 4.0 initiatives[1]. The fields of statistics, chemometrics, and machine learning are expected to provide tools that effectively handle data characteristics, such as process measurements (e.g. temperature, pressure), spectra from online analyzers and quality control labs, pictures taken with standard or hyper-spectral cameras, and all other data sources that provide insight into the state of product processes and quality.

The task of image classification[2] is considered in this paper. A supervised learning problem is presented whereby an image is the input and the output is a unique label attributed to the image from a set of available classes. Image classification is one of the main tools used for quality assurance and control at Dow Chemical, and developing a suitable classifier is a relevant industrial challenge due to accuracy and robustness requirements. In this context, recent developments in deep learning[3] that have proven successful in increasing image classification accuracy and providing state-of-the-art results in computer vision. Traditional approaches utilized for image classification are based on prior knowledge and pre-defined features. However, in this work, we leverage deep neural networksâ€™ (DNN) ability to automatically learn features from images[4] and test their performance in a real industrial context of pellet shape prediction. Moreover, other less complex techniques such as partial least squares discriminant analysis (PLS-DA) and random forests (RF) are explored in order to assess the benefits of adopting DNN as opposed to a more traditional classifier.

PLS-DA, RF, and DNN models were developed for two classification tasks: pellet shape classification (distinguishing good and bad pellets), and detecting tails in a pellet (distinguishing whether a pellet contains tails or not). After developing these models, the results were consistent for both classification objectives. Compared to the in situ classification system currently used at Dow Chemical, RF were able to better utilize the same pre-defined features and improve prediction accuracy significantly, while PLS-DA had the worst performance. DNN obtained the highest accuracy overall (higher than 96%), and the main advantage is that there is no need to specify a priori features because they are extracted from the raw image itself. Furthermore, visualizing the output of some layers of the network showed that activations occurred in regions that are meaningful for the classification tasks, further supporting that DNN effectively modeled the relevant features of the pellet.

References

1. Reis, M. and G. Gins, *Industrial Process Monitoring in the Big Data/Industry 4.0 Era: from Detection, to Diagnosis, to Prognosis.* Processes, 2017. **5**(3): p. 35.
2. Duchesne, C., J.J. Liu, and J.F. MacGregor, *Multivariate image analysis in the process industries: A review.* Chemometrics and Intelligent Laboratory Systems, 2012. **117**(Supplement C): p. 116-128.
3. LeCun, Y., Y. Bengio, and G. Hinton, *Deep learning.* Nature, 2015. **521**(7553): p. 436-444.
4. Simonyan, K. and A. Zisserman, *Very deep convolutional networks for large-scale image recognition.* arXiv preprint arXiv:1409.1556, 2014.

513668 Immersive Introduction to Iot

Abstract:

ABSTRACT for AIChE 2018 Spring Meeting

Immersive Introduction to IoT

Joel Berg

The MathWorks, Inc.

joel.berg@mathworks.com

508-647-4415

You've heard of the "Internet of Things" (IoT), and have a notion of it, but what is it really? And how many presentations do you attend where the speaker actually asks you to turn on and use your phone?!

This talk will begin with a short overview of some common building blocks in IoT architecture, including sensing and measurement devices, wireless communications, cloud-based data aggregation, real-time model updating, and decision/action rules taken by cyber or physical actors.

Now the fun starts, with the presentation becoming more of a group "event." We will describe a process-oriented scenario, and ask audience members with smartphones to participate as actors within a living, cyber-physical model. Dynamic information will be shown and continuously updated on the main screen, and participants will play active roles according to signals they are seeing displayed and a brief set of instructions, by way of their devices (smartphones). The aggregated inputs of the audience will govern the system state evolution, which will also be displayed. The event will end with a short discussion and inventory of observations and insights deriving from the demonstration.

The objectives of this talk/event are to educate attendees about IoT, introduce the ThingSpeak® platform, provoke discussions about IoT use cases, and, last but not least, have some fun.

512435 Comparing Several Methods for Product Properties Prediction

Abstract:

Comparing several methods for product properties prediction

Jean Jérôme Da Costa[1], Benoit Celse[1], Fabien Chainet[1], Marion Lacoue Nègre[1], Cyril Ruckebush[2], Didier Espinat[1],*

1. *IFP Energies Nouvelles, Rond-Point de l'Echangeur de Solaize, 69360 Solaize, France*
2. *LASIR CNRS, Université Lille Nord de France, Sciences et Technologies, Cité scientifique, bât C5, 59665 Villeneuve d'Ascq Cedex - France*

**E-mail: benoit.celse@ifpen.fr*

HIGHLIGHTS

- Comparison between linear regression models and interpolation methods on toy examples
- Interpolation methods are much more efficient especially for non linear data
- Comparison between kriging and spline interpolation
- Kriging models can provide a posteriori analysis of the results
- Application on cloud point and viscosity index prediction

INTRODUCTION

In order to respect specification, petroleum products properties prediction is an increasingly need for refining industry. Machine learning is then more and more used to develop predictive models from process and physicochemical analytical data. The most classical models are based on classical (multilinear and nonlinear) regression referring to some equations which describe physicochemical phenomena or derived from empirical observations (Riazi 2005). However, regression models require to identify the adapted analytical expression for the regression function. That may be very complex in some situations.

Many chemometrics applications in modeling petroleum products properties were also proposed using partial least squares (PLS) regression (Braga, Jez Willian B., Junior, Araci Araújo Dos Santos et Martins 2014). Although these models provide relatively good performances, they use spectroscopic data such as NMR (nuclear magnetic resonance) or NIR (Near Infrared spectroscopy). This type of data are not currently available for refiners.

Some authors (Piloto-Rodríguez et al. 2013) proposed to use artificial neural networks (ANNs) for modeling complex properties. The results are good. However, the optimization of ANN requires to perform very stringent statistical steps to insure the model robustness and a lot of effort to prevent from over-fitting. This might be very tricky.

The aim of this work is to compare interpolation and linear regression methods on toy and real examples (could point, viscosity index ...) in order to suggest the best methodology to use for product properties prediction.

MATERIALS & METHODS

Kriging is an interpolation method based on stochastic multiGaussian assumptions (Roustant, Ginsbourger et Deville 2012)). Its goal is to provide the best linear unbiased estimate (B.L.U.E.) and was first used in applied geostatistics for modeling properties related to natural resources. More recently, other applications of kriging were developed such that metamodeling, which is an increasingly need in computing sciences (DiceKriging, DiceOptim: Two R Packages for the Analysis of Computer Experiments by Kriging-Based Metamodeling and Optimization 2012). One major point of kriging is the ability to provide a measure of the prediction uncertainty that depends on the data configuration (Roustant, Ginsbourger et Deville 2012).

The aim of splines interpolation is to draw contour lines "as smooth as possible", that is a map which looks like what a draftsman would obtain manually (Rooij, P. L. J. van et Schurer 1973). The splines consist of polynomials of degree m being local. The polynomials describe pieces of a line or surface. For degree m=1,2 or 3, a spline is called linear, quadratic or cubic respectively. This method is currently used in non-parametric statistical learning (Schumaker 2015).

RESULTS & DISCUSSIONS

Results on toy examples point out that interpolation methods have a high capacity of adaptability to linear or nonlinear situations without requiring previous analysis. Although, kriging and splines provide very close performance in some situations, kriging is more valuable since it provides a measure of uncertainties related to the predicted value based on stochastic assumptions. It is based on a specific distance between the new sample to predict and the calibration data base. This point is very important in modeling physicochemical properties of petroleum products. It is even particularly essential when the number of descriptors increases significantly (higher than 3).

Kriging and MLR models are developed and compared in real examples: cloud points and viscosity index. Although MLR provides good performances that are in accordance with the uncertainties of the standard measure, kriging enables to improve accuracy. In our knowledge, this is the first kriging model for the prediction of these two properties.

Globally, interpolation methods are well adapted to the modeling of petroleum products properties since they provide good performances when the number of samples is limited. Use of kriging methods to high dimensional study remains always challenging.

Bibliography

Braga, Jez Willian B.; Junior, Araci Araújo Dos Santos; Martins, Ingrid S. (2014) Determination of viscosity index in lubricant oils by infrared spectroscopy and PLSR. In : Fuel, vol. 120, p. 171–178. DOI: 10.1016/j.fuel.2013.12.017.

DiceKriging, DiceOptim: Two R Packages for the Analysis of Computer Experiments by Kriging-Based Metamodeling and Optimization (2012) : University of California, Los Angeles.

Piloto-Rodríguez, Ramón; Sánchez-Borroto, Yisel; Lapuerta, Magin; Goyos-Pérez, Leonardo; Verhelst, Sebastian (2013) Prediction of the cetane number of biodiesel using artificial neural networks and multiple linear regression. In : Global Conference on Renewable energy and Energy Efficiency for Desert Regions 2011 "GCREEDER 2011", vol. 65, n° Supplement C, p. 255–261. DOI: 10.1016/j.enconman.2012.07.023.

Riazi, M. R. (2005) Characterization and properties of petroleum fractions. 1st ed (Online-Ausg.). West Conshohocken, Pa : ASTM International (ASTM manual series MNL 50).

Rooij, P. L. J. van; Schurer, F. (1973) A bibliography on spline functions. Eindhoven : Technische Hogeschool (TH report / Technische Hogeschool Eindhoven Nederland, 73-WSK-01).

Roustant, Olivier; Ginsbourger, David; Deville, Yves (2012) DiceKriging, DiceOptim: Two R Packages for the Analysis of Computer Experiments by Kriging-Based Metamodeling and Optimization. In : Journal of statstical software, vol. 51, n° 1.

Schumaker, Larry L. (2015) Spline functions. Computational methods. Philadelphia : Society for Industrial and Applied Mathematics (Other titles in applied mathematics, 142).

AUTHOR INDEX